# Credit Where Credit Is Due: Respecting Authorship and Intellectual Property

ACS SYMPOSIUM SERIES **1291**

# Credit Where Credit Is Due: Respecting Authorship and Intellectual Property

**Patricia Ann Mabrouk**, Editor
*Department of Chemistry and Chemical Biology*
*Northeastern University*
*Boston, Massachusetts*

**Judith N. Currano**, Editor
*Chemistry Library*
*University of Pennsylvania*
*Philadelphia, Pennsylvania*

**Sponsored by the**
**ACS Division of Chemical Information**

American Chemical Society, Washington, DC

Distributed in print by Oxford University Press

**Library of Congress Cataloging-in-Publication Data**

Names: Write Thing to Do: Ethical Considerations in Authorship & the Assignment of Credit (Conference) (2017 : San Francisco, Calif.) | Mabrouk, Patricia Ann, editor. | Currano, Judith N., editor. | American Chemical Society. Division of Chemical Information, sponsoring body. Title: Credit where credit is due : respecting authorship and intellectual property / Patricia Ann Mabrouk, Editor, Department of Chemistry and Chemical Biology Northeastern University Boston, Massachusetts; Judith N. Currano, Editor, Chemistry Library, University of Pennsylvania, Philadelphia, Pennsylvania ; sponsored by the ACS Division of Chemical Information, American Chemical Society, Washington DC. Description: Washington DC : American Chemical Society, 2018. | Series: ACS symposium series ; 1291 | "This volume is based on the symposium "The Write Thing to Do: Ethical Considerations in Authorship & the Assignment of Credit" held at the 253rd National Meeting of the American Chemical Society, which took place in San Francisco, CA on April 3, 2017." | Includes bibliographical references and index.

Identifiers: LCCN 2018026505 (print) | LCCN 2018030011 (ebook) | ISBN 9780841233386 (ebook) | ISBN 9780841233393

Subjects: LCSH: Authors and publishers--Congresses. | Scholarly publishing--Law and legislation--Congresses. | Authorship--Collaboration--Congresses.

Classification: LCC K1488.A6 (ebook) | LCC K1488.A6 W75 2017 (print) | DDC 346.04/82--dc23

LC record available at https://lccn.loc.gov/2018026505

The paper used in this publication meets the minimum requirements of American National Standard for Information Sciences—Permanence of Paper for Printed Library Materials, ANSI Z39.48n1984.



Distributed in print by Oxford University Press

# Foreword

The ACS Symposium Series was first published in 1974 to provide a mechanism for publishing symposia quickly in book form. The purpose of the series is to publish timely, comprehensive books developed from the ACS sponsored symposia based on current scientific research. Occasionally, books are developed from symposia sponsored by other organizations when the topic is of keen interest to the chemistry audience.

Before agreeing to publish a book, the proposed table of contents is reviewed for appropriate and comprehensive coverage and for interest to the audience. Some papers may be excluded to better focus the book; others may be added to provide comprehensiveness. When appropriate, overview or introductory chapters are added. Drafts of chapters are peer-reviewed prior to final acceptance or rejection, and manuscripts are prepared in camera-ready format.

As a rule, only original research papers and original review papers are included in the volumes. Verbatim reproductions of previous published papers are not accepted.

**ACS Books Department**

# Contents

## Authorship in Collaborative Research Projects



## Indexes

# Preface

Participation in undergraduate research played a formative role in my development as a young scientist, but it was authorship that was the linchpin in the development of my identity as a scientist. Many years ago, before active learning methods or classroom undergraduate research experiences were known, as an undergraduate I participated in a course where the professor leveraged the class to collect data, which he subsequently published as a research study. My name and that of all my classmates were listed in the byline of the paper but I only discovered this years later when doing a google search. I never saw a copy of the manuscript, so I have never listed this paper on my CV.

The summer before I began graduate school, I began doing research in my doctoral advisor's laboratory. Due to the novelty of our work, I was extremely fortunate that my doctoral advisor sought to publish my early findings, and my first paper was published in *Chemical Physics Letters* during my second year of graduate school. All the data and figures, traced laboriously using India ink on tracing paper from the paper-based spectral traces produced by the instrument I helped build, were mine but most of the words were not (lots of red on the many drafts of that short paper!). At that point, I had much to learn about science and the effective communication of research, but that authorship experience was pivotal in the development of my professional self-identity. It was the first time I saw the bigger picture - how my work and I fit into the greater scientific community of practice. I have always marveled at the paradigm shift that occurred through my first experience with authorship. Consequently, I have sought to provide my students an opportunity to engage fully in the realization of their research scholarship.

## Vision

This volume is based on the symposium "The Write Thing to Do: Ethical Considerations in Authorship & the Assignment of Credit" held at the 253$^{rd}$ National Meeting of the American Chemical Society, which took place in San Francisco, CA on April 3, 2017. Both editors, serving on the ACS Committee on Ethics, felt that there was a need for more focused, in-depth resources on critical ethical issues, such as assignment of credit. Our goal has been to develop a robust resource that explores the central issues from a variety of perspectives within the greater chemical community of practice encouraging a healthy discussion of the key issues related to assignment of credit including authorship, contributor-ship, inventorship, and copyright. As educators, we envisioned the book as a scholarly resource that college and university faculty teaching research ethics to postdocs, graduate students, and undergraduates could use both in the classroom and in their group meetings as well. So, we sought out Jeffrey Kovac and John D'Angelo, authors of two well respected textbooks on science ethics to craft case studies on the issues discussed in each chapter of this book.

## Limitations

We learned a lot about authorship during the process of putting this book together. Originally, we intended to include chapters exploring authorship credit in private industry and from a global perspective but ironically, differing cultural values on authorship precluded the participation of several individuals. We also intended to include the voice of an important independent organization, the Committee on Publication Ethics, but concerns about digital publication rights led this organization to withdraw from our project.

## Final Words

While no book can perfectly capture all of the voices, ideas, and facets of subjects as complex as authorship and the assignment of credit, we hope that we have made a good start and one that promotes a healthy discussion. We truly welcome your feedback on this book and your ideas for next steps in expanding this conversation.

**Patricia Ann Mabrouk**
Department of Chemistry and Chemical Biology
Northeastern University
Boston, Massachusetts 02115

# Acknowledgments

As you might surmise from the preface, my journey to realize this book has had a few unexpected twists and turns that have led me to a deep gratitude for everyone with whom I partnered including the chapter authors, reviewers, and the ACS and Technica professional staff. Truly this project would not have been realized if it were not for the talent and generosity of the chapter authors, many of whom were dealing with serious challenges during this project and yet powered through their circumstances to fulfill their commitment: Thank you Art, Eric, and Judith! Many thanks to Tracey Glazener at Technica Editorial for all of her help navigating the day-to-day process required to realize this book and her unending patience dealing with me. I cannot thank Sara Tenney at ACS Books enough for listening, ready assistance, and grace in dealing with often challenging partners. Heartfelt thanks go to Kaitlin McKernan, Charles Nino, Andrew Min, Akira Nishii, and Brian Coppola for their reviews of the case studies in this book; if the case studies are successful in promoting student learning, it is due to your thoughtful and insightful feedback. Lastly, heartfelt thanks and my sincerest admiration go to my co-editor on this project, Judith Currano, for sharing my vision and coming through under adversity.

**Patricia Ann Mabrouk**
Department of Chemistry and Chemical Biology
Northeastern University
Boston, Massachusetts 02115

# Acknowledgments

Many thanks to all of the talented individuals whose hard work made this book a reality. I would especially like to thank Peter Allen for helping me to find the time and energy to see the project to completion.

**Judith C. Currano**
Head, Chemistry Library
University of Pennsylvania
Philadelphia, Pennsylvania 19104

# Assigning Credit: Defining Authorship, Contributorship, Inventorship, and Copyright

## Chapter 1

# Assigning Credit and Ensuring Accountability

## An Editor's Perspective on Authorship

**Keith S. Taber***

**Faculty of Education, University of Cambridge, 184 Hills Road, Cambridge, England, CB2 8PQ, United Kingdom**
***E-mail: kst24@cam.ac.uk.**

Academic authorship is a key concept in scholarly publication. Publications bring academic credit, and authorship is the accepted way of recognizing who deserves that credit. Similarly, authorship ensures accountability for the claims published in research journals. Journals, therefore, commonly require the submitting author to make some form of declaration that the submitted manuscript includes an accurate author list. Academic authorship is not focused on the process of writing a scholarly article, but on the intellectual work which is reported in that article. Generally academic works offer knowledge claims, which are considered useful contributions when judged as novel and well supported by argument and evidence. The research paper is therefore an account of an argument based on considerable work undertaken prior to the writing process, and authors are those who have contributed substantial intellectual work to the study. All named authors need to approve the submitted manuscript, as they will be considered to share accountability for it if published, but it is not necessary for all authors to substantially contribute to the writing as long as they have contributed to the thinking behind it. Even though the criteria for academic authorship are straightforward, there is much scope for dispute in interpreting the principles. The chapter discusses the nature of academic authorship and the possibility for inadvertent and deliberate errors in assigning authorship. The chapter also suggests some simple guidelines to minimize the potential for problems to develop over questions of authorship.

## Introduction

Academic authorship is a key concept in scholarly publication. Publications bring academic credit, and authorship is the accepted way of recognizing who deserves that credit. Someone named as an author who made no contribution, or no substantive contribution to the work reported is being awarded credit that is not deserved and that should be assigned elsewhere. Someone who made substantive intellectual contributions to the work reported but who is not named as an author is not able to support a claim for the academic credit that their contribution to the work deserves. Journals, therefore, have a responsibility to acknowledge the contributions of all those deserving to be seen as authors of works. Similarly, authorship ensures accountability for the claims published in research journals. Publishers and editors cannot vouch for the accuracy or honesty of published studies, and peer review can only offer a certain level of quality assurance, so authorship implies that a scholar is prepared to put their professional reputation behind their published work. Journals, therefore, commonly require the submitting author to make some form of declaration that the submitted manuscript includes an accurate author list.

Yet, despite the centrality of authorship claims in academic life, academic authorship is not always well understood, even among the academic community itself, as some of the examples discussed later in the chapter will demonstrate. This has the potential to lead to misunderstandings, academic disputes, and even suggestions of malpractice or unprofessional behavior. A more challenging issue is that, even when the concept is understood, the criteria for academic authorship need to be interpreted when deciding who is an author of a particular manuscript. This means that it is quite possible (and indeed happens in practice) that sometimes colleagues working on a project may have the same principled basis for understanding what counts as academic authorship, yet they still disagree on who should be considered authors of particular outputs. It is in the nature of interpretation that it has a subjective component (there may be a tendency for individuals to have a distorted view of their own contributions to collective work), so, this is not something that can ever be completely avoided. However, this chapter will offer some simple guidelines for minimizing opportunities for conflicts to arise over who deserves authorship on particular studies.

This chapter will discuss how journals expect those submitting articles to understand the notion of authorship and why determination of authorship is seen as an ethical issue when publishing scholarly work. The chapter explains why authorship can be a problematic issue and offers guidance to the community that is relevant to individual scholars; those having a role as editors; those having responsibility for the research training, academic induction, and supervision or mentoring of research students and new academics; and potentially of those in senior positions who may be faced with mediating a dispute between junior colleagues in their departments.

## The Concept of Academic Authorship

In everyday discourse, an author is the person who wrote something; that is, the author formulated the text. There is a widespread understanding that authorship is not simply the mechanical process of producing a manuscript, but, in most everyday contexts, it is certainly tied to the creation *of a text*. The writer who dictates her novel to a secretary is the author of the work, not the person who takes down dictation or types up the manuscript. A text is a form of artistic creation, deriving from the imagination of the author.

When considering scholarly, rather than artistic, works, there is still a text that has been created (sometimes with considerable craft and literary skill), but the interest of readers is less on that text *qua* text in its own right than on the scholarly content it reports. These are, in effect, claims to knowledge (this point is developed in more detail below). Both a literary text and a scholarly text are protected by copyright; that is, an author has rights over a text that allows them to permit or prevent publication or, indeed, to license or sell on those rights for some consideration, as well as the right to be named as the author of the work if it is published. Ideas are not subject to copyright; only the text itself is. If there are figures or tables, then the ideas or data they are meant to communicate are not protected by copyright, but the designs themselves are.

As scholarly texts are primarily appreciated for their arguments for knowledge claims, rather than as works of art, being recognized as the originator of a new idea or claim is important (*1*), and authors of academic works are usually more concerned with having their ideas published (under their names) than with having their texts protected from wider publication. Journal editors are also likely to be more interested in authorship in relation to the development of ideas presented in manuscripts, and this is the main concern of this chapter, but they must also consider the legal requirement not to publish material that infringes copyright because permission has not been given by the legal copyright holder. Copyright is usually initially held by an author, although some writing undertaken as part of employment may be considered work-for-hire. The precise arrangements may vary according to contracts of appointment or local laws, but generally a formal document (such as a policy document or an examination paper) written by an academic for their department as part of their duties will be considered copyright of the institution, although individual researchers will hold the copyright in their scholarly work.

## An Academic Text as an Argument for New Knowledge Claims

In the empirical sciences, such as chemistry and chemical education, texts are not primarily valued as texts (that is, as a particular literary construction), but as representations of ideas. A research report will offer knowledge claims, which will be supported by a line of argument based on the interpretation of data as supporting evidence (*2*). A review article that revisits a canon of existing studies is similar, but with the previously published works taking the role of data for the review. A theoretical contribution, such as a perspective article, is somewhat different but also makes knowledge claims of a kind.

Research reports in chemistry make claims about natural phenomena, and those in chemistry education make claims about behavioral or social phenomena. It is generally recognized that empirical research rarely makes direct claims about the unmediated nature of the world. In what might be called a post-positivist age (*3*), it is acknowledged that scientists (in natural sciences, such as chemistry, or in social sciences, such as education) are developing theories, building models, establishing typologies offering first-order classifications, identifying laws that may only refer to non-existent ideal states, and so forth, rather than establishing true and absolute descriptions of the nature of things.

Scientific ideas are therefore creative: they are imagined possibilities about how things might be (*4*). Those worth publishing in reports of empirical studies must offer imagined possibilities that can be shown to be broadly consistent with the data collected to test them. In a well known aphorism, "essentially, all models are wrong, but some are useful" (*5*). That is, although we should not mistake our models and theories for direct accounts of how things are, and although sometimes an acknowledged degree of mismatch with available evidence may be tolerated, we are ultimately seeking to imagine how things are (unlike in artistic work), not simply to imagine how they might have been.

Theoretical contributions, such as perspectives, also offer new ways of imagining some aspect of the world, perhaps an aspect of the material world or perhaps an aspect of the social world, either in terms of ontology (how things are) or epistemology (how we might find out how things are). However, for such contributions to be considered worth publishing, they must be seen as novel and as potentially productive in terms of supporting a progressive research programme, whilst also being linked to the established canon of ideas guiding the field (*6*). The set of ideas synthesized or developed in the study, which has a role parallel to empirical data in this kind of article, may be drawn from the field (*e.g.,* chemistry education), may borrow from another field or fields (*e.g.,* from anthropology into chemistry education), or may draw upon methodological literature (*e.g.,* techniques from neuroscience not previously applied in the field). A research programme is considered progressive (*7*) as long as it is growing through the addition of productive new empirical or theoretical content, but a new theoretical perspective will only add to the programme if it offers guidance on directions for further empirical work (that is, it supports what is referred to as the programme's positive heuristic). Figure 1 offers a schematic representation of the scientific paper, understood as a logical chain where different kinds of knowledge claims are developed from working with different kinds of 'data'.

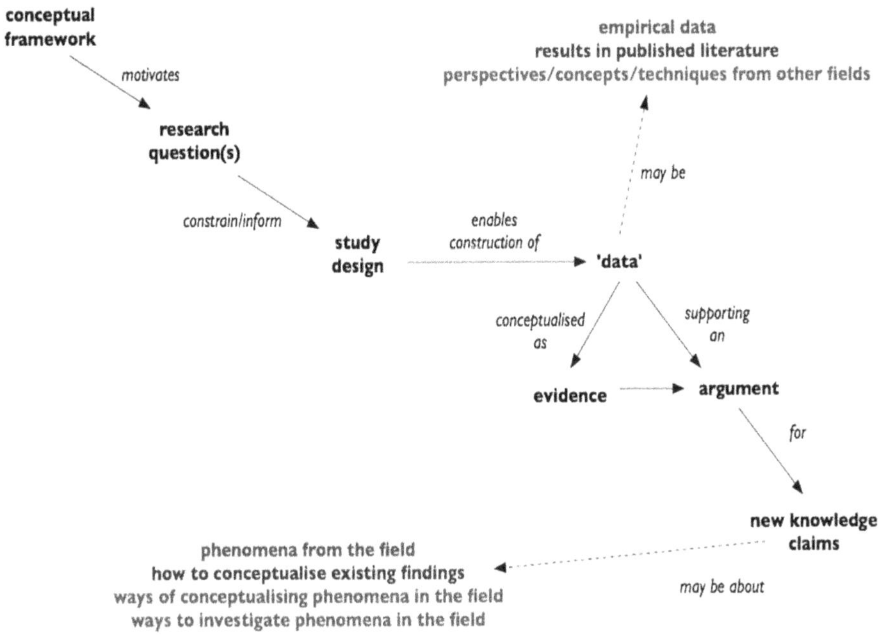


*Figure 1. Articles in scientific journals progress a research programme by making supported claims for new knowledge.*

## Academic Authorship as Intellectual Work

Authorship of academic articles is, then, not so much tied to the paper as a text in its own right, but more to the production of the knowledge claims represented in it. The creative product in an academic, rather than a literary, context is a supported argument for a knowledge claim. An author is someone who contributes substantially to that production. So, in relation to any published paper, the authors should be those who have made substantial intellectual contributions to the work being reported (not just the text itself), and only those who have made such contributions.

One leading chemistry education journal has stated that:

> *"An author of an academic work is someone who made a substantial intellectual contribution to the work [so therefore] ... (a) all those who made substantial intellectual inputs to a work should be named as authors; (b) only those who made substantial intellectual inputs to a work should be named as authors (8)."*

So, someone should not be named as an author of a scholarly work unless they have actually contributed to its production. Moreover, that contribution needs to have been substantive rather than peripheral, and should be an intellectual contribution rather than, say, practical assistance or logistical support. Although this principle appears deceptively straightforward, there are three ways that problems can arise in relation to assigning authorship of academic works (see also Figure 2):

- Individuals may not understand the criteria that are usually applied to determining academic authorship (they perhaps hold an alternative conception);
- Different individuals who do understand the criteria that are usually applied to determining academic authorship disagree on how to apply these criteria in a particular case (there are differences of interpretation);
- Sometimes individuals who do understand the criteria that are usually applied to determining academic authorship deliberately misrepresent authorship of academic work (*i.e.,* they behave unprofessionally and unethically).

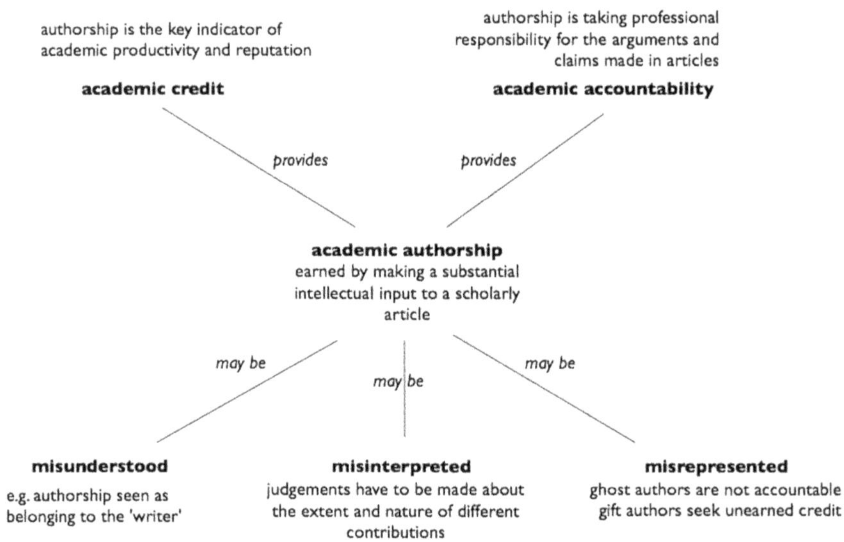


*Figure 2. Key issues of authorship.*

In principle, the third possibility reflects a different kind of problem, as it involves deliberate, rather than inadvertent, disregard for academic norms. However, like any typology, these distinctions model a more complex situation, and it is likely that "custom and practice" within particular cultural and institutional contexts may sometimes lead to behavior that is best understood as somewhat intermediate between situations 1 and 3 or between situations 2 and 3. For example, it may have become routine that a head of institution or group is named as an author of outputs, such that the custom continues even when it is no longer justified by the level of engagement in particular studies; this may happen without anyone consciously intending to misrepresent authorship.

The crux of the issue is that academic authorship is highly significant, so, it is important that authorship of published works is correctly assigned, and it is problematic if deserving researchers are excluded from author lists or if undeserving researchers are included. As suggested below, journal editors have good reason to suspect submitted manuscripts do sometimes have inappropriate author assignments.

## The Importance of Being Accountable

In parallel with the prestige that comes with acknowledged authorship comes responsibility. Just as authors deserve credit for their contributions (discussed further below), they need to take responsibility for their work (*9*) and therefore offer accountability for faults in the work (*10*). Many chemistry education researchers will have come across Student's T-test that is often used to compare similarity in means of data samples (*11*), but may not realize that Student was an assumed name used by the statistician W. S. Gossett at the request of his employers so that his true identity would remain unknown; such an arrangement would not usually be acceptable to journals today.

As Susser (*12*) expresses the matter: "a key assumption of the tradition is responsibility for what one publishes, and hence accountability for what is false, fraudulent, or taken without acknowledgment". Journals, therefore, commonly expect the submitting author will not only assure the accuracy of the author list, but will confirm that all those named as authors have approved the final form of the text submitted. The text presents the knowledge claims made and the argument that supports them, and being named as an author implies one is prepared to stand by both the claims and the argument.

### Authorship Entails Responsibility to the Scholarly Community

The literature in a field is cumulative. A field moves forward not through individual studies as isolated achievements, but through the iterative progress made when researchers build upon what has gone before (*7, 13*). Some knowledge claims are more productive than others in this sense. Many published papers are seldom cited. Others are regularly cited for many years after publication, and some may attain the status of "classics", but this is not predictable at the time of publication. Even published papers that never come to be cited may inform the work of academics and graduate students, given the phenomenon of publication bias: in other words, that "at each stage of this process [*i.e.*, of undertaking, writing up, submitting, peer reviewing, and editorial decisions on publication] the probability of progressing to the next stage is strongly influenced by the results of the analysis, with, in general, a significant result being a positive inducement to proceed, as opposed to a non-significant result" (*14*). A reasonable *prima facie* conjecture here would be that published reports with untrustworthy findings are more likely to lead to work that (fails to offer positive findings and so) is then not considered worthy of publication itself.

The status of historical scientific literature can be seen as analogous to the history of the biota on the earth. The vast majority of species that have ever existed became extinct, and those that still exist today will likely all become extinct at some point in the future. Similarly, most published research papers have been succeeded by others, which show they were mistaken or which reconceptualize, extend, or refine their findings. Empirical studies that might be considered extant, in the sense that researchers still refer to them for guidance in carrying out new studies, are like those species that are extant: the rare exceptions. Most, probably all, papers considered to reflect the current state of knowledge today will also be surpassed in the future. In a sense, then, all scientific papers come in time to be seen as flawed, even those that are highly influential. (Classic scientific papers tend to be of much more interest to historians and philosophers of science than to those working in the source scientific field today.)

However, papers can be flawed in different ways. Consumers of scientific articles have a right to expect they are honest reports of careful work. A study that is an honest report of work that has been carefully carried out may prove to be flawed for a number of different reasons. In particular, flaws may primarily be matters of conceptualization, of procedure, and of logic.

**Distinguishing Potential Flaws in Published Papers**

What is meant by a flaw of logic is a poor chain of argument, such that, even on its terms, the paper does not make its case. In other words, even if someone accepts a paper's conceptual framework as a fair starting point, the interpretation of the data collected does not convince the reader that the knowledge claims are sufficiently supported. This might be because the reader feels that the research design is insufficient, the sample is too small or is not representative of the intended population, the data are insufficient, the analytical techniques are inappropriate or insufficiently rigorous, or the interpretation of the analyzed data does not support the conclusions drawn.

Flaws of this kind are often spotted in peer review, so, ideally, such papers should not enter the literature. If they slip through the process, then a reader can make his or her own judgement. Others may judge that the work does not make a productive contribution, but the authors have done their best, worked carefully, and reported honestly. This should not harm the field, although it is worth noting that the recent explosion of new open-access journals with dubious peer review standards (*15*), which require authors to pay for publication, increases the proportion of published peer-reviewed articles that an expert might consider seriously flawed. This is a concern in a field like chemistry education, where relevant studies may draw upon a wide range of perspectives and methodologies (*16*), most of which are outside the expertise of any particular scholar in the field.

Another kind of flaw concerns conceptualization. It may be judged that the paper's starting point is an understanding of the existing state of the field that is not supported by the literature, the research questions are posed in terms of a theoretical framework that does not usefully illuminate the topic of the paper, or something of that nature. The paper makes a logical argument for someone who accepts its premises, but those premises are not accepted by (some) readers;

therefore the conclusions cannot be relied upon. This case divides into three types of situations.

One category concerns the situation in which others working in the field judge the conceptualization inappropriate at the time that the work is done. In this situation, the paper should be judged as flawed in peer review, and either the authors should demonstrate that they can make changes to preserve the logic of the study, or the work should not be published (bearing in mind the caveat above about the uneven quality of peer review). A second category concerns genuine differences of opinion among the community of researchers in the field, such that some referees would accept a particular conceptualization as productive that others would not find viable. Such papers are sometimes published (perhaps not in the first journal to which the paper was submitted), and researchers must then make their own judgements on how convincing they find claims made in such papers. A third category consists of papers that are conceptualized in ways that seem acceptable and productive at the time of their submission, but that will later (perhaps soon after publication, perhaps years or decades later) come to be judged to be poorly conceptualized. Scholarship moves on, and even the work of icons, such as Newton, Dalton, and Darwin, is conceptualized in ways that would now be questioned. In the social sciences, the work of highly influential authors, including Marx, Freud, or Piaget, has been subject to extensive critique but remains influential for opening avenues of scholarship. In natural sciences, literature reviews are expected to be more "up to date".

In some of these cases, academics will commit time and other valuable resources to carry out work, in part because of published findings, conclusions, and recommendations in work that may later come to be seen as flawed. However, if the original work was carefully undertaken and honestly reported, this has to be seen as part of the ebb and flow of the progressive nature of science, and no blame accrues to the original authors, who did their best to offer an accurate account of their work that others could judge for themselves.

Other flaws are matters of procedure. These may be technical or human faults. Power supplies may fluctuate. Materials may not have the purity claimed by suppliers. In education, published instruments supposedly demonstrated to be valid and reliable may be found not to be so. Computer software used to analyze data may have programming bugs. Researchers need to be on their guard against such matters, but sometimes glitches will arise that are not detected at the time despite due diligence. These glitches may well misdirect authors, and so, potentially, their readers. *Caveat emptor*: the consumer of research needs to be aware that these things can happen.

Human beings make errors. They misclassify. They miss or repeat data when entering it for analysis. They make comments before starting an interview that inadvertently channel a study participant to think in a particular way. They use leading questions without realizing they are doing so. In idiographic work that uses semi-structured interviewing, it is a matter of careful, "in-the-moment" judgment how one phrases questions, how-open ended one's questions are, when one uses follow-ups or changes a sequence of questions, etc., and such issues are seldom clear-cut. Most serious errors can be detected and corrected by employing study procedures that include sufficient checks. Occasionally, though, errors may be missed. Again, this should be appreciated by readers but should not be frequent.

In these cases, published work may misdirect future researchers or practitioners who base their practice on the evidence offered in research papers. However, when such mistakes lead to flawed conclusions, it is likely that such a paper will be noticeably at odds with other similar studies, thus alerting the community to be wary of its claims. Careful and honest papers are sometimes based on mistakes, but we can be tolerant of the occasional honest error.

Shoddy or dishonest work is less acceptable. Work that has been carelessly conducted is likely to be untrustworthy, but this may not be obvious in a carefully-crafted account. A study that uses substandard procedures but is deliberately written in a way that hides this may appear much more convincing than the actual work it (mis)reports. For example, if researchers review their interviews of participants and notice that they have been using leading questioning, they need to acknowledge and allow for this when using the data as evidence in the development of arguments to support new knowledge claims. Yet, it is possible to ignore the issue (especially given what peer reviewers might think about such an admission) and to carefully select data snippets that do not reveal the flaw in order to illustrate the findings. In such situations, readers may invest resources in carrying out research or introducing teaching innovations based on what is, in effect, falsified work. This, perhaps, does not happen very often, but when it does, it can mislead the community. Yet, that is avoidable and only occurs because some authors prioritize simply getting published, publishing notable findings, or publishing findings supportive of their preferred theoretical position over offering an honest and trustworthy account. This is clearly unethical, but the pressures to publish are so great that there have been calls for revisiting the academic norms that make having publications, and a lot of them, such an imperative for career progression (*12*, *17*).

Because of the very nature of reports that misrepresent shoddy work, these papers may go unspotted. This makes it especially important that authors who are recognized as falling short of what is expected are asked to take responsibility for misleading the community. For example, as a reviewer, I discovered one paper submitted for publication, in which the authors had misrepresented the sample size by multiplying by four the number of participants taking part and the numbers of individuals responding in various ways. Normally, there is no way a reviewer could know if a sample reported to be 80 students was actually only 20 students. It was just chance that I had previously reviewed "the same" (or, rather, virtually the same) manuscript for another journal; the two manuscripts were almost identical word for word but not number for number.

Readers may also assume that researchers are competent in the techniques that they use. An illuminating example is randomization. Many statistical techniques only offer interpretable results when comparing two conditions if there has been a randomization process. Selecting a random sample of a large population, for example, of all high school chemistry teachers in the United States, is actually a challenging technical task. Yet, assigning one of two classes, or one of two teachers (an approach which is weak, in any case), to one of two conditions randomly only calls for a very simple technique of the type used by referees or umpires before sports matches when deciding which side gets to choose "ends", who should kick off, or which side will bat first. Despite this, on more than one occasion, I have found that authors of submitted manuscripts that include claims that they assigned teachers or classes to conditions randomly are later completely unable to inform me how they did this. This has occurred frequently enough to establish an expectation at the journal *Chemistry Education Research and Practice* that, whenever a paper reports random assignment, it needs to be accompanied by a brief description of the means used (*18*). Surprisingly, perhaps, it is not always safe for a reader to assume that researchers in chemistry education know the difference between a random assignment and an arbitrary assignment.

## Ghost Authors

It is important, then, that those substantially responsible for the intellectual work of generating knowledge claims reported in scholarly works are identified so that they can take responsibility for that work. It might seem unlikely that a deserving author would freely consent to not being named on a submission, but, in some fields of research, there has been a concern that actual authors are sometimes absent from the author lists on submitted papers. These are referred to as ghost authors: "when someone has made substantial contributions to writing a manuscript and this role is not mentioned in the manuscript itself" (*9*). Such ghost authors cannot be held responsible for their claims or the quality of work on which such claims are based.

This is a different situation to that in which the contributions of junior researchers (or people having left a research group) are inappropriately ignored by principal investigators when submitting a paper, a situation that is described later in this chapter. Ghost authors choose to contribute without acknowledgement, working in a manner akin to those ghost writers of celebrity books who are happy to take a fee (or perhaps a share of the royalties) and to have their names appear in small print in a book's front matter, rather than being prominently displayed on the cover and spine. Academic ghost authors are not acknowledged at all, however. This is of concern because of potential conflicts of interest.

The worry is that, in areas like drug studies, the pharmaceutical industry may employ professional science writers to do the writing for researchers. This ensures that studies important to the industry are published, while leaving the researchers more time for carrying out their research. This practice is of concern because, even if the named researcher is able to be fully objective, the ghost writer is not independent, and this fact is hidden from the journal editors and reviewers, as well as from the readers of the published study. The researcher, as the named author, takes responsibility for the study (and gets full credit for the work), but it is not clear how careful they are in checking, and, if necessary, modifying the text provided by the ghost writer (*19*). It seems unlikely that ghost authorship is currently a major issue in chemistry education, but it is important for the community to realize that the omission from author lists of those who should be authors is unacceptable even when (or perhaps especially when) those concerned do not wish to be named.

**Plagiarism and Copyright Infringement**

Some thinkers have sought to argue that any writer's actual text necessarily has multiple, or at least distributed, authorship. The argument is that the writer has to be seen within a wider cultural context that provides the resources for writing (*20*). The final text might be produced by an individual, but it borrows phrases, ideas, style, and so forth from the cultural milieu. In scholarly writing, the author is expected to acknowledge the sources of thinking derived from the published work of others. To fail to do this is plagiarism. Yet, in practice, the evaluation of plagiarism is much more nuanced. Any author can inadvertently plagiarize ideas from work once read but not directly remembered. Some people are better at remembering the original form and details of the works they read than others, but, by its nature, the human brain is very effective at adapting, re-interpreting, and melding ideas. It seems to have evolved to be effective at offering a continuously-updated and largely coherent model of the world as experienced, rather than a carefully-distinguished and catalogued account of discrete past experiences (*21*). "Photographic" memory seems to be rare among adults, and when it does occur, it is not necessarily an advantage (*22*).

Moreover, all authors use ideas of other thinkers that they indirectly acquired from culture without the origins being explicit. Perhaps readers of this volume are aware of the origins of such notions and idioms as "the collective unconscious", "the medium is the message", "big brother", "the usual suspects", "crossing the Rubicon", "I am Spartacus", etc., but, in common discourse, many people use these references without any awareness of their literary or cultural origins.

The steps necessary to avoid suspicion of plagiarism vary from field to field. In the natural sciences, there is seldom an expectation to cite the origin of well-accepted scientific principles which are taken to have become the shared property of the community. Dalton, the Lavoisiers, or, indeed, Pauling, are seldom cited in contemporary research reports that inherently rely upon ideas they brought into chemistry. Usually, only the more directly proximate precursors of the reported study are expected to be cited. Whilst adopting this approach is seen as good practice in the natural sciences, it would likely be judged poor scholarship in the humanities. In science, offering an expansive list of references including classic works considered to have been long since surpassed would be considered inappropriate. In some other scholarly fields *not* citing such works would be considered just as inappropriate.

Chemistry education, as a field, does not have well established norms in this regard. In peer review, manuscripts may be criticized for not referencing originators of lines of research or conversely for reference lists that cite too many "outdated" studies. Sensibly, a balance is needed, paying due reference to classic studies but acknowledging more recent work that has moved the research programme forward. The expectation to cite prior work is sometimes interpreted by authors as offering a literature review consisting of a series of general summary statements, each followed by long lists of relevant studies. Reviewers may criticize such statements as both being too indiscriminate and offering insufficient explanation of those contributions that genuinely deserve to be acknowledged. These are matters of scholarly judgement (seemingly somewhat influenced by different cultural norms) that can generally be addressed by effective peer review processes.

Substantive copyright infringements of textual material are likely rare (and many journals have access to software tools to check for this), with the possible exception of authors' own works, considered below. It is quite common, however, for manuscripts submitted for publication to include graphics copied from other texts. It is sometimes considered that citing a source is sufficient here because authors do not consider the distinction between plagiarism and copyright infringement. To scan a figure from a book or download it from a website and include it in a manuscript avoids *plagiarism* if the original source is cited, but the original design is likely to have a copyright owner (the original designer, or a publisher) and cannot be copied legally without permission. The copyright owner is entitled to refuse permission or to ask for a fee. Permission is often forthcoming and, when an image derives from another academic source, will often be granted without charge, but some publishers are much quicker than others in responding to requests. In submitting a manuscript to a journal, the author(s) declare they are the copyright owners or otherwise have received permission to reproduce work of any third-party copyright owners, and they may be responsible for financial damages if a journal publisher is sued for unauthorized use of copyright material. Copying a figure (or a table) by hand is an infringement of copyright in the design as much as scanning or photographing it. *Re*drawing the figure (so that the design is changed, sufficiently, such that the authors could convince a court it is not merely a copy) avoids needing permission, and then citation is sufficient to avoid accusations of plagiarism.

In principle, texts are protected, as well as designs, and a sentence in a book or article is subject to copyright. In practice, fair use exemptions are allowed for modest direct quotations from most scholarly works, and they may be used for purposes of scholarship and critique with citation but without needing permissions. However, this does not apply to all artistic works, so quoting just a few lines from a poem still in copyright is likely to need express permission.

One area where these issues commonly become problematic is that of multiple publication. Again, it is important to consider both duplicate publication (sometimes considered "self-plagiarism") and copyright as related but distinct issues. It is not generally acceptable to publish the same paper in several places, although there may be exceptions. Authors are often allowed to republish their journal articles (with suitable acknowledgements to the original publication) in anthologies of their work. Occasionally, a classic paper may be republished with permission (perhaps with various commentaries) some years later. A paper might also be published in translation in another language. Generally, though, publication of an article in one journal precludes its publication elsewhere.

Historically a paper published in an obscure, inaccessible journal might later be published in an international journal, although in the "internet age" such a level of inaccessibility is probably an anachronism. For example, in 1989, Peterson, Treagust, and Garnet published a paper about the development and findings from a diagnostic instrument on covalent bonding and structure in the US-based *Journal of Research in Science Teaching* (*23*), which was, in effect, a longer version of a paper previously published in *Research in Science Education* (*24*), then a journal largely restricted to readers in Australasia and primarily publishing conference papers. Now, both journals are considered leading international journals and are largely accessed through the Internet.

It might seem unlikely that an author could inadvertently seek to republish, but I have known this to happen. It was noticed that a paper submitted for publication in a major journal had already been published in a relatively unknown open-access journal. In this case, it appeared that the author was unaware of the existing publication, as she had declined to pay a publication fee that she had been told was a condition for having the article published. Presumably, a license to publish had already been signed, and it had not seemed necessary to withdraw it.

A more common problem is one of overlap between what are clearly different but related articles. Publishing much the same study with minor variations is not acceptable. Clearly what is considered 'minor' variation is open to interpretation, but if a submission clearly acknowledges overlap with existing work, then reviewers and editors can reach a view on the extent of novelty of the submission, as long as the relevant related publications are cited. As an example, Taber and Tan (*25*) published a paper comparing the responses of pre-service chemistry teachers and the responses of senior high school chemistry students to a diagnostic instrument related to ionization energy. The paper revisited material previously published as two discrete studies (*26*, *27*), and offered a new cross-study analysis. It was necessary to include previously-published findings (in effect, now data for the new study), but care was taken to be explicit about precisely which material was previously published so that reviewers of the manuscript, and later readers of the published study, could be quite clear about the nature of the novel contribution of the new article.

Publishing versions of work for different audiences should not be problematic, as the natures of the different accounts should be clearly distinct: for example, publishing a research report in a journal with technical details and then an account aimed at practitioners focused on implications for practice (*28*), which cites the research report.

Publishing a range of studies focusing on different aspects of the same project is certainly justified, and, again, peer reviewers can reach judgements on whether particular submissions are rich enough to stand alone as a study (*cf.* Figure 1). In terms of plagiarism, a reviewer or editor would expect to see distinct differences in the research questions, findings, and implications in the different articles, even if they are drawn from the same project. If two reports are coming to the same conclusions from the same data sets in response to the same research questions, then they can reasonably be seen as comprising the same study and should be reported in a single article. However, in large, complex research projects, it often makes sense to report the project in discrete chunks that allow each paper to focus on a particular aspect. This will better allow a clear line of argument to be developed (*cf.* Figure 1).

Authors should be careful of directly re-using text from one submission in another manuscript, particularly whole sections of literature review and methodology. Although a study's specific claims and supporting argument are the aspects that make it novel, the authors will likely have already signed away or licensed copyright to the text that they may be tempted to re-use. There is no copyright applied to ideas reflected in a conceptual framework or a research design, so these may be described repeatedly, subject to an acknowledgement of where they were already published, but once text (a particular form of words) has been published, the author may not be entitled to use it again without the express permission of a publisher.

Authors need to be aware of the rights they retain and those they sign away, when submitting work for publication. The precise form of words used in a previously-published study probably cannot be used again without permission of the publisher unless the work was published under a license that specifies such use is allowed. If it was published under such a license, then the author can use the text again but cannot assign copyright in it to the new publisher (which tends to be requested for a publication that will not be open-access). The sensible advice to authors is that, if some text has already been published, one should not use it again without substantial modification unless it is clear one has retained the right to do so. The situation may get even more complicated when co-authored studies are the source material. It is an interesting experience as a reviewer to realize you are reading text in a blinded manuscript that you originally crafted.

## The Importance of Recognizing Contributions

Authorship is of central importance in determining an academic's scholarly reputation and her or his career prospects; an author gets credit for publications, as well as being publicly responsible for them. Evaluations made in relation to appointment, tenure (in those countries where this concept operates) or completion of probation, and promotion are often in large part influenced by the candidate's publication list: that is, their claim of academic authorship over scholarly works. Such lists are expected to include particular types of "output", such as refereed articles in academic journals, scholarly books, chapters in edited volumes, and so forth. Such lists may also feature as part of the required evidence used to evaluate such matters as requests for visiting scholar status in other institutions, proposals for support from funding bodies, or nominations for academic prizes and awards. In many universities, an academic's work is judged in relation to three areas: research, teaching, and a general contribution to the administration and management of the department and university and to the wider academic community (through involvement in journals, work with organizations like the American Chemical Society (ACS) or the Royal Society of Chemistry (RSC), or through work in public engagement, etc.). This general contribution is also sometimes referred to as "service".

In appointments to research-intensive universities, the potential to contribute through research (primarily judged in terms of publications: how many, where published, how often cited, etc.) is often considered to be the most critical factor in reaching such decisions, even when contribution through teaching is seen as an equally important aspect of the role. Senior academic promotions, such as appointments to personal chairs, will depend upon evaluations of contributions to all three areas of academic work but may quite deliberately and explicitly put a premium on contributions through research, in effect research publications, for which academic authorship will often be the most visible indicator of academic productivity.

The principle that *an author of an academic work is someone who made a substantial intellectual contribution to the work* is a fairly simple idea (if not always simple to apply, as discussed below). However, even this basic principle is not universally recognized in the scholarly community, as I found when I invited someone to review a submitted paper in an area of work in which they seemed to have particular expertise. I found that they had even more specific expertise for evaluating that paper than I could have expected. The colleague declined to review, on the grounds that they should have been considered an author of the paper. When asked, the submitting author did not deny that the work being reported was undertaken by a team of researchers. However, one of these had been a visitor in the group who had since returned home, and the others were graduate students working on their theses. The submitting author seemed genuinely upset that the students had not had time to contribute to writing the paper, and so the text was produced by a single writer. As far as this individual was concerned, authoring a paper was synonymous with writing it, but that is not the understanding expected when submitting work to journals.

Work undertaken by graduate students working under academic supervisors is worthy of particular consideration. A potentially confusing variable across systems is the nature of the Ph.D. thesis. In some contexts, it is expected that the student will be embedded in a research group, and will help develop a strand of research from within the group's joint research programme that will lead to a series of outputs, such as conference papers and journal articles. These will be co-authored, but the student will be expected to be the lead author on at least some (this system supports a kind of academic apprenticeship wherein the student takes increasingly central roles on particular studies). The thesis, in the student's name, becomes a kind of edited volume, with introductory and concluding sections written by the student that bookend a series of chapters comprising co-authored papers, some of which may have already been published.

A rather different tradition expects the student to write a monograph, a book-length dissertation as a thesis, for which the student is the sole author. This may be framed as a single study or as a series of related studies, but it is presented as 'all my own work' by the student. The doctoral candidate is expected to make a declaration to that extent, and detail any support obtained in preparing the thesis - for example if a peer helped by repeating some data analysis for a reliability test. Such theses often acknowledge help from various sources, but help that is not presented as worthy of authorship.

Occasionally a doctoral student may complete their dissertation without any support at a level that would normally comprise co-authorship: but this is probably rare. The normal assumption in a doctoral program is that the new research student requires close supervision and induction into a research field. That support includes guidance on conceptualization of a project and research design, and very commonly a supervisor makes extensive suggestions for reworking draft chapters. Even if a student arrives with a well-considered research idea, this is likely to be significantly modified in conversations with the supervisor(s) and/or other advisors; and sometimes the final project is essentially set out by a supervisor for the student. That is, the student's dissertation is treated as if a sole-authored document, even though it would normally be seen as a co-authored work in terms of the expectations of academic publishing.

This may confuse students, who may either assume that any work reported in their dissertation that is submitted for publication can rightly be considered single authored; or worry that if they submit articles from the doctoral project to journals as co-authored (whether before or after the award of their Ph.D.) then this undermines the claim to the dissertation being all their own work required by the University.

## Matters of Interpretation

Such problems should be addressed by education: any one induced into a field of scholarship should be introduced into its norms, conventions, and expectations (*29*): whether this be related to matters of conceptual commitments, research design, instrumentation, statistical conventions (*i.e.,* the usual choice of $p \leq 0.05$ as a criterion for significance) or modes of dissemination. When journals in a field expect authorship to reflect all substantive scholarly contributions to the work reported, it is not acceptable that anyone should complete their research training in the field without knowing this.

However, whatever definition is used to define authorship, it will be open to interpretation, as given the diversity of scholarly studies such a definition will necessarily be vouched in generalities. Taking the definition presented earlier, it has to be decided what constitutes *substantial intellectual input*. 'Substantial' clearly suggests that one does not assign authorship on the basis of a minor or incidental contribution. Someone who suggests a relevant study to be included in the literature review of a paper, or even that a particular perspective or research technique (which is taken up) might be useful, does not make a substantial contribution even if their suggestion substantially affects the direction of the study.

"Why don't you try a think-aloud technique" could inform a critical feature of research design, but making a suggestion is insufficient for authorship. There is a suitable means for assigning credit for a useful suggestion, and that is to include an acknowledgement. However, someone who suggests that a think-aloud protocol might be useful, and is then asked for, and gives, extensive advice on the nature of the technique, and how to incorporate it within a research design, and advises on the development of the protocol and the analysis of the data, might well be considered to deserve authorship. They have perhaps only given advice (rather than carried out the data collection and analysis), but that advice might amount to a major contribution to the design and trustworthiness of the study.

The other component of the guidance given for authors of submission to *Chemistry Education Research and Practice* is that a contribution must be not only substantial but also intellectual. This means that contributions to data collection and analysis that are substantial in terms of the commitment of time (perhaps amounting to a fair proportion of the total hours of work undertaken on the study) may not necessarily be considered sufficient for authorship.

We might consider here a common distinction between studies which are deductive and confirmatory in nature and seek nomothetic knowledge, and those which are inductive and exploratory in nature, and often more focused on idiographic knowledge (*16*). For example, consider a study which used an oral questionnaire with standardized questions to carry out a survey compared with a case study where participants were interviewed based on a flexible open-ended interview schedule. We can imagine that in both cases a researcher was employed to help with data collection and analysis, and for argument's sake spent 100 hours undertaking the assigned tasks.

In the first hypothetical study, the researcher asks questions, exactly as provided, and marks responses into pre-determined categories, before entering the data into a computer database to be analyzed by a statistical program. If this researcher had no role in designing or developing the questionnaire, the sampling frame, or the statistical techniques to be used, then although their input was substantial in terms of time, it is unlikely to amount to a substantial intellectual contribution. It would seem inappropriate to offer this researcher authorship for this level of contribution. Certainly their contribution was necessary, but any other suitable qualified researcher could have been hired to do the work to the same level.

In the second study, a researcher can only be given instructions on how to carry out the work of data collection and analysis to a limited degree. Many in-the-moment decisions need to be made about the interview process: when to rephrase questions for participants; when to reflect answers back for confirmation, or use follow-ups to develop a line of response; when to drop, or to re-sequence, questions, in the light of comments made; when to introduce an additional line of questioning triggered by something unexpected in a participant's comments; when to push for further answers, or instead sense a participant's reluctance or fatigue, and to step back. Unlike when using a set of standard questions, a researcher can only effectively interview in this modality if they have a good understanding of the background and purpose of the research as this will inform many of the myriad decisions that need to be made. The researcher needs to understand the research context, and to be able to develop a rapport with the individual study participants. This kind of interviewing cannot be purely a technical contribution.

Transcribing interviews in such a study is a highly interpretive task rather than a largely mechanical one. Analyzing interview transcripts is skilled work requiring immersion in the data, and again needs to be informed by a strong understanding of the rationale of the study. Unlike in quantitative analysis, when the logic of an enquiry needs to be built into the design of the instrument, construction of a sample, and choice of statistical tests, and so the analysis itself is an algorithmic process that can be quickly and effectively carried out by a machine (*i.e.,* because the 'thinking' part of the analysis is actually carried out before the data is collected), in interpretative research a unique human analyst with relevant interpretive resources (understanding of the conceptual framework, of the study context, of the participants) is needed to do the job.

Data collection and analysis in such a study are then more than just technical tasks, and part of the trustworthiness of the study will depend upon the skills and engagement the researcher brings to the work. This researcher is surely making a contribution to the task which is both substantial, and comprises intellectual work. These hypothetical examples show that judgments about authorship need to be taken in the context of particular studies.

### How Many Chemical Educators Does It Take To Author a Study?

Chemistry education is a field where many papers published are single authored, others are authored by pairs of scholars, and others are authored by small research teams. In general though, published studies seldom have more than a handful of authors. In some areas of the natural sciences, there has been concern how in recent decades papers have started to appear which have very long authorship lists. In part this phenomenon links to the development of so-called 'big science' (*12*). Some areas of research depend upon extensive collaborations.

The most high profile example is high energy ('particle') physics. Results in such a field may depend upon multinational funding, and consortiums of research teams. Moreover, although the knowledge claims are made in the discipline of physics, they rely upon teams of engineers, statisticians, and computer programmers to help carry out the physics. That is, the arguments for the physics knowledge claims rely upon instrumentation and analytical tools developed in cognate fields. Knorr Cetina (*30*) has offered a fascinating account of how results are constructed in such a field showing how the actual data collected by detectors is made meaningful though a complex apparatus of computer simulations and statistical analyses. Given that results may also be so dependent upon enormous grants negotiated through organizations comprising many national and institutional interests, it may be appreciated that authorship may not only be difficult to determine, but also subject to political considerations. The results are papers with dozens, indeed hundreds of authors.

A paper attributed to the 'ATLAS collaboration' published in *Physics Letters B* (*31*), took up about twelve pages of journal space to just give a full list of the contributors and their institutional affiliations. With such long teams of authors, it has now become normal practice to only actually list a limited number of those authors when the papers are cited. To make a point, the full list of authors was included in the reference list in one instance when the paper was cited, and that single reference entry occupied 9 pages of the article (*32*) - the space needed to list all of 2932 authors. Indeed there were so many contributors considered worthy of some of the credit for this study that 21 of the authors listed as members of the collaboration had died by the time the paper was published.

This situation raises questions about the traditional understanding of authorship. As suggested above authorship both acknowledges credit, and ensures there is accountability and responsibility for error. But if a study with over two thousand authors were found to be problematic, it may be hard for the academic community to know who actually 'is Spartacus'.

To date this has not become an issue in chemistry education, but has been noted as a concern in some other fields, including areas of medicine. Trends towards longer author lists in some fields have invited the question "How many neurosurgeons / cardiac surgeons/ orthopedic surgeons does it take to write a research article?" (*33–35*). There is certainly a suggestion that sometimes those offered credit as authors on research papers are being given this as a gift or consideration unrelated to the extent of their contribution to the work being reported.

### Guest Authors

So-called guest authors, or honorary authors, those not making a substantive intellectual contribution to a study, may be offered authorship for a range of reasons. Twenty years ago Susser claimed:

> *"Authorship has been extorted as the price of needed data or even patients. Honorary authorship has appeased heads of departments or secured the cover of their prestige. Gifts of authorship have sustained friendships or merely marked kindness toward a colleague or junior in need (12)."*

Scientists working alone have been known to name pets as co-authors to avoid the use of the first person in their writing (*32*). Apparently it has sometimes been considered more scientific and trustworthy to write 'we' rather than 'I', even if the co-author is a dog or cat who presumably made little intellectual or practical contribution to the work. Of course, the co-author is not identified as an Afghan hound or a Siamese cat at the point of submission, but rather simply named as Galadriel Mirkwood (*36*) or F. D. C. (*i.e.,* Felis domesticus Chester) Willard (*37*).

Perhaps one of the most well known examples of guest authorship is the so-called α,β,γ paper (*38*) where George Gamow added his friend and colleague Hans Bethe to the author list of a paper deriving from his graduate student Ralphe Alpher's doctoral thesis work, supposedly out of respect for the Greek alphabet. Given the importance assigned to authorship today, it seems unlikely that such whimsy, or indeed a reluctance to write in the first person, can be considered acceptable grounds for sharing authorship in this way. It is worth emphasizing that these articles supposedly coauthored by Mirkwood, Willard, and Bethe, were not provocative opinion pieces but serious scientific studies. At the time of writing this chapter Willard's paper had 77 citations on Google scholar; and Mirkwood's 130. The Alpher, Bethe and Gamow paper had been cited 1067 times. A good many scholars never achieve anything like a thousand citations for any of their publications, let alone for a paper for which they made no genuine contribution.

The example of particle physics certainly shows the need for particular fields to develop their own norms and conventions. It has been suggested that in some biological fields it is accepted that when research relies upon the use of particular materials, such as a certain cell line, a colleague providing that material should be accredited as an author (*30*). This type of contribution would not be expected to deserve authorship in the physical sciences. However, within a particular field it may be recognized that research not only relies upon the availability of specific (*e.g.*, biological) material, but that the production and quality assurance of that material is a highly specialized matter. It may therefore come to be accepted in the field that there is a genuine collaboration such that authorship is deserved in a way which would not be considered when purchasing readily available materials from a commercial supplier. The question that may be raised (*12*), is whether such authorship represents a judgement that a genuine intellectual contribution to the new study has been made, rather than a judgement that the work is not possible without such offering of authorship for material.

In chemistry education research, a study might make use of materials such as data collection instruments used in previous publications by other authors. These may be used as originally published, in translation, or even adapted for a new context. Citation to the original source is expected, and if permission is needed to use the instrument (or if the instrument is not itself published and is only available on request from the original investigators), then a specific acknowledgement may be justified. However, providing an existing instrument is not usually considered as making a substantive intellectual contribution to the further study, and authorship should not be offered for the use of the instrument.

Unlike a cell-line whose viability needs to be maintained, providing a copy of an existing instrument such as a questionnaire does not involve special skills (i.e. beyond what is already credited through the original publication). Indeed, granting permission for the use of such an instrument in exchange for authorship would be inappropriate and unethical. However, if the investigators of the new study not only ask to use the instrument, but engage in such extensive correspondence or direct conversation with the developer(s) of the instrument that it amounts to a genuine input into the conceptualization and design of the study, then authorship may well be justified and deserved.

Similar principles can be applied in other situations. The term 'publication parasitism' has been used to describe the situation where a junior researcher is pressured to name a senior researcher as an author on their papers when in their judgement that person has not made a significant contribution to the work (*10*). The Head of an institution, lab, or research group should not expect to be automatically named as an author on outputs from their unit on the basis that they provide employment, funding, facilities or a buffer to higher authorities such as deans or funding agencies; nor because they are responsible for staff evaluation or providing references. However, if that Head develops a research programme, and offers an overall conceptualization within which projects within the unit are developed, and if that head is familiar with each study, and regularly engages in providing significant guidance and feedback to the investigators on each study, then this could well amount to authorship. No amount of bureaucratic or political support justifies authorship, but genuine engagement in projects that offers scholarly leadership can amount to significant intellectual input even when such engagement reflects a limited time commitment. Once again, as interpretation is involved (what counts are genuine engagements in intensive research-leading conversations, rather than polite enquiries about progress), judgements have to be made.

A related issue is the so-called 'publication cartel' where a group of researchers mutually gift authorship to each other. Collaborations where several researchers engage in genuine dialogue to inform each other's research may lead to synergism, and so be productive, and could justify genuine co-authorship. However to agree to '*I'll name you if you name me*' simply as a strategy to artificially enhance publications lists is a form of malpractice - defrauding the community by misrepresenting the authorship of studies. A system of high stakes research assessment that evaluates institutions depending in part upon the proportion of academics regularly publishing (such as found in the UK) has potential to distort authorship decisions as departments only get credit for a limited number of publications per 'returned' academic, which can lead to pressure for the most productive scholars to share some of their work, so more of it can be included by being returned under their 'co-authors'.

### Ordering the Authors

Where work is genuinely co-authored, there is a need to consider the order in which the names should appear. In some fields there are already conventions for this (*30*), but these may be misunderstood outside those contexts. In some parts of the world it is common to include names in reverse order of seniority, but this is by no means universal. A sensible convention is to always seek to order the names according to contribution to the work.

This may not be straightforward: revisiting the discussion earlier of what counts as a substantive intellectual contribution, the most important contributions may not imply the greatest amount of time spent on the work. Sometimes strategic oversight at an 'executive' level (*8*), perhaps involving just a few hours per week, counts for much more in determining the nature and quality of a study than full-time day-to-day execution of the research plan.

In the case of papers deriving directly from students' own projects, it would seem sensible to expect that when co-authorship is justified (see above), the student will be likely to have made the most substantial contributions to the intellectual work, and so should normally be named as the first author. Certainly if this is not the case it should reflect a clear agreement between the authors, taking care that the student is not showing too much deference, either out of their respect for their supervisor, or concern for the hierarchical structure in which they are placed. It is a key principle of academia (even if one that is difficult to ensure in some circumstances) that all work should be judged on its own merits without regard to reputation or standing of its author(s). The person denoted as the corresponding author of a paper (the person readers are directed to engage with in correspondence about the work) need not be the first author, nor indeed the submitting author, so it is perfectly possible for the student to be first author, and the supervisor the corresponding author - although again this should be decided through a conversation between the authors.

## Balancing Considerations When Determining Authorship

It has been suggested that the principles involved in determining academic authorship are straightforward in principle, even if they become complicated in practice. However there may be occasions when expectations regarding the ethics of authorship seem to conflict with other value-informed considerations, or even when ethical considerations conflict and so do not seem to offer an unambiguous choice of course of action. This section will briefly consider such issues.

## A Responsibility To Support Junior Researchers

Authorship should not be gifted, but must be earned. Yet this may seem harsh when a research team includes junior researchers, such as (but not limited to) graduate students, as their careers will likely be dependent upon building up a publications list. The senior investigator may have a formal development role in relation to the junior researcher, or may simply feel a moral responsibility to help them as much as possible.

The argument here is that this is very worthy, but that authorship should not be assigned when it is not justified. As discussed above, there may sometimes be questions of interpretation where it is debatable whether a contribution does or does not deserve authorship: if there is genuine uncertainty then risking erring on the side of generosity seems a more honorable choice than risking excluding someone with a legitimate claim to authorship. That said, wherever possible the potential for such ambiguities should be avoided (see the concluding comments). If a senior academic feels a responsibility to develop a junior colleague, this needs to be done by giving them a sufficient role within a project such that authorship is earned, not gifted.

### *Graduate Students and Research Supervisors*

The situation regarding graduate students is complicated by different norms in different national contexts. So a doctoral candidate may be either considered as a student paying fees (either personally, or through a scholarship) for supervision, and who is expected to focus primarily on a thesis project; or alternatively in effect an employee of the university who is expected to split their time between a dissertation project and supporting research and/or teaching in their department. In the latter case, undertaking some work as a research assistant, possibly in a largely technical capacity (i.e., perhaps not at a level amounting to authorship), is expected in return for providing facilities and support of the doctoral project.

In the former situation, the student would have limited time for working on other projects, and if invited by a supervisor to contribute to other work this should clearly be when the opportunity is valuable as a training/development experience. In either case, best practice would be to seek to engage the student so they are a full part of a research team - both so they learn from having an overview of the full logic of the research (cf. Figure 1), and so that there is potential to contribute at a level commensurate with being named as an author. Even if what is needed from the graduate student or other junior colleagues is technical level input, it seems irresponsible to limit the researcher's contribution in this way if their involvement is meant to have a developmental role. It would seem more sensible to also involve them closely in other stages of the work: in discussions on research design, building a sample, selecting an instrument, interpreting findings, etc. Although a junior partner may have less to offer, they will learn by being involved and if they engage in the process there is a much stronger case that authorship is appropriate.

## The Role of a Translator

Issues that may be less obvious concern contributions made by those who are not usually considered part of the research team. One issue concerns translation. Most of the top international scholarly journals publish their articles in English. Authors who are not native English speakers may have different levels of proficiency in forming English text. Such authors may reasonably seek assistance in refining their writing.

Instruments used to collect data, such as questionnaires or teaching materials, and data that may need to be presented in articles, such as extracts from interviews, student work, etc, will often not be in English, but will need to be presented in English. Here there is more than an issue of tidying up an author's own translation, but a need to offer readers an assurance of the trustworthiness of material presented in translation.

Where a translator is called upon, this raises an issue of whether authorship is involved. In most cases of simply refining an author's text, this may be a favor provided by a colleague or a service provided by a commercial organization. In this situation input is normally best considered technical, worthy of acknowledgement but not sufficient for authorship. However, if there is a need to translate substantive material, such as interview texts, where the basis of the claims made in the article rely upon this evidence, and where a trustworthy translation relies upon scholarly knowledge of the field and the specific study, as well as of the two languages, it may be that such a contribution begins to approach authorship. A decision that this is not authorship certainly should be one deliberately made, and not simply assumed.

Each language offers a unique set of resources for describing and indeed constructing the world and translation is not a mechanical process of substituting 'the equivalent' word from another language: "the translator must take the liberties of an author to subvert language in order to transfer a literary work into the cultural and textual context of the target language" (*39*). Idioms may be highly idiosyncratic and culturally sensitive. The issue is more extreme with literary works, but also applies in more technical contexts. The credit given to translators varies considerably, and the creative input of the translator to the translated text (i.e., as something akin to co-authorship) is recognized much more in some national contexts than others (*39*).

## The Unheard Voice

Another group who are not usually considered part of the research team but who are essential to the research in education are participants - such as teachers or students. Again here it is useful to separate out issues relating to intellectual input and copyright. In many research studies participants offer the gift of data (*40*). If this includes samples of student or teacher writing, or their diagrams, then these would likely be considered to be their copyright (*41*) and researchers need to be sure that voluntary informed consent sought covers permission to use such material in publications (*42*). The usual expectations of citation of sources are complicated here by the normal practice of offering anonymity to participants.

This type of contribution would not normally amount to a claim on authorship of the resulting report. Yet it is important to be aware that in some types of study (such as a case study, for example) where an individual contributes substantial text, this could be challenged. Moreover, issues of relative status and power may be at work in determining whether originators of extensive source material are recognized as authors. Where the professional ghost writer may be happy to shape a celebrity's book without seeking prominence as an author, writers who have told the stories of those with limited cultural capital - such as the illiterate poor, or indigenous people from non-literate cultures - may or may not feel they need to share authorship with those who provide the stories they tell (*43*). This may not simply be a matter of the writer being in a position of sufficient power to allow them to ignore the informant's claims of authorship, but a question of habitus (*44*): a lack of awareness on the part of the informant that there is even an issue of authorship to be considered (and indeed an issue which may be linked to achieving status and financial rewards).

Potentially a similar situation may occur when a researcher collects extensive data from one learner, or even from a teacher, who may well not have given thought to such issues as authorship of the outcome of such a collaboration. If the researcher feels it is appropriate to share credit, for example, when the participant is the 'subject' of a case study that amounts to collaborative work, there may then be a clash between the desire to offer deserved credit through authorship and an assurance that the participant's identity will not be revealed. Including the participant as an author under the assumed name used in the research (*45*) is one solution but is clearly not entirely satisfactory.

## Challenges in Determining Authorship

This chapter has suggested some of the problems that arise in determining academic authorship of articles submitted for publication. The expectations of most journals are now so clear that instances of guest authorship and ghost authorship should not occur by accident. Despite this, fairly recent research shows that, in some fields, the incidence may still be significant: it has been estimated that a fifth of all published papers in high-impact biomedical journals have author lists that are either incomplete or include people having made insufficient contributions to deserve authorship (*46*). This was based on a questionnaire sent to corresponding authors, which might have led to underreporting in the survey.

Where there is potential for dispute is over matters of interpretation, it is helpful for expectations to be established at the start of any research collaboration. An invitation to undertake work on a research project should make it clear if what is wanted is considered purely a technical contribution perhaps worthy of acknowledgement but not authorship, or is intended to amount to being part of the author team taking strategic responsibility for the work (and being accountable for any account later published). In complex studies, different outputs may reasonably have different authorship lists from within the wider team, and the ordering of authors may vary reflecting different levels of contribution to different parts of the overall work. Again it is important to set out expectations, and to engage in conversations when (as will inevitably sometimes happen) any member of the team thinks these need to change. Ultimately no one should be surprised at the point when a submission is prepared by whether they are a named author, or where they are placed in the author list. Discussion of authorship early in a project may seem a little distasteful (like discussing money issues between friends) but just as misunderstandings might strain friendships when the waiter brings the bill at the end of a meal, it is sensible to ensure no one is surprised and upset by being excluded or listed last at the point where an article is submitted.

No set of guidelines can however address all the situations that might arise. I am aware of one project where the investigators came to disagree over how some of the results should be presented. Some of the team thought there were strong reportable results, but others felt those conclusions were somewhat undermined by some of the data collected as a check on the degree of similarity of the different conditions being compared. This could not be resolved as neither side could be persuaded by the other. Some of the team refused to be named as authors on a submission unless certain information was clearly discussed: information that the other investigators did not feel it was relevant to highlight in the report.

Ethical guidelines did not offer a satisfactory outcome in this situation. Those who felt the paper was flawed and misleading would not agree to its submission without modifications that were not acceptable to the others. The options in such a situation seem to be:

1. Submit the article, and name all those who should be considered authors (although not all thought they could stand behind the knowledge claims made in the form reported) - satisfying credit, and asking some to take on accountability for a report they could not stand by;
2. Submit the article with a reduced list of authors (even though those omitted had made contributions to the work that should have amounted to authorship) - fitting the notion of accountability, but not fully reflecting credit; or
3. Not publish - so no account of the work would be available.

At first sight, option 3 may seem to be the most ethical option as it avoids contravening expectations about authorship. However, when research has involved a considerable use of precious resources, especially when it is publicly funded, and involves a large number of participants inevitably inconvenienced in facilitating the work, there is also an ethical imperative to publish.

In the end, option 2 was adopted. Two of the investigators refused to have their names on the submission unless it was modified, but the rest of the team went ahead. The potential limitations of the study that were in dispute were reported in another more generic publication from the same research project but were not brought to the attention of readers in the disputed article: something that was, or was not, problematic depending on the judgments of different members of the research team. Those named as authors felt they were prepared to be accountable for its contents, but others who made substantive intellectual contributions to the work (but would not stand behind the claims made in the particular output) were merely acknowledged for their input. This was not an entirely satisfactory outcome, but the only fully acceptable outcome would have been to publish a version of the article all could sign off on, and it was judged by those involved that this was not going to happen. The outcome was not satisfactory, but was the best compromise that those involved could find.

It is difficult to know how often such situations arise, and how they are then resolved. It seems unlikely any work would ever be published in fields like high energy physics if each author had to be persuaded of the merits of each aspect of the report without relying on the expertise of others. By contrast in a field like philosophy, work is nearly always single authored, given that the author of a philosophical work is taken to be responsible for the precise formulation of language in which arguments and claims are made [see Sainz, in reference (*32*)]. Chemistry education is a field which occupies a somewhat intermediate position.

## Conclusion - Good Practice in Managing Authorship in Research

Journals expect submissions to have a full and accurate authorship list, and the submitting author is expected to affirm this, and also that all those named as authors have approved the text for submission. Journals generally expect the acknowledgement of support from funders, and clear statements of any potential conflicts of interest. Some journals now ask for details concerning which authors contributed to which aspects of a study, either on all studies or those with a large number of authors. Where such details are made public, then accountability for different aspects of a work may be distributed rather than collective. However, commonly it is assumed that submitting authors understand and are following expected conventions, and such details are not requested.

The discussion of authorship issues in this chapter leads to two basic guidelines that can avoid or solve most (if sadly not all) potential disputes over authorship. These relate to education and the early contractual agreement on expected research outcomes. Firstly, it should be an expectation that there will be explicit discussion of the nature of authorship as part of doctoral education. Research students should be taught about the academic notion of authorship, preferably as part of the doctoral research training course so that students have a chance to explore the idea within the supportive context of their peer group. Supervisors should be expected to raise this subject early in the supervision relationship so that an expectation is established as a basis for conversations that will need to take place later. It is certainly not the case that any article submitted to a journal by a research student should be automatically considered co-authored: but rather that the student should learn that any academic writing that derives from a collaboration (which effective doctoral supervision should be) requires a conversation between the collaborators about whether the particular piece of writing should be considered co-authored, and if not whether an acknowledgement of support is appropriate instead.

The declarations made at the submission of the dissertation, commonly require the students to declare the dissertation is their own work, except as detailed. This may sometimes be interpreted by the education student as an expectation that a dissertation derives from the intellectual work of a sole scholar, when in some disciplines (including much work in natural sciences) such a dissertation would be very rare. Again education is needed to prepare the doctoral candidate to set out in their acknowledgement the level and degree of support obtained from supervisor(s), advisor(s) and others. A clear statement of the involvement of the supervisor 'covers' the student in terms of the declaration, and supports the rationale for including work which may be submitted for publication as co-authored. A statement that provides a detailed account of the division of labor, is good preparation for academic work.

There is usually no reason why material reported in a dissertation should not be included in a journal submission, nor why work published or being considered for publication should be excluded from a dissertation, although in either situation appropriate acknowledgements should be made. A dissertation should include acknowledgement of which material is published, or is submitted for potential publication, at the point of submitting the dissertation for examination. Journal articles drawn from a published dissertation should cite the dissertation and acknowledge the source. Articles submitted from a dissertation under preparation should include an acknowledgement that the work reported derives from an ongoing graduate project, indicating the institution and department and the principal supervisor (whether listed as an author or not). Although priority disputes (arguments about who first reported some result or proposed some idea) are not common in chemistry education, it is in the interest of the author to ensure that a journal article acknowledges when work reported has appeared in a dissertation with an earlier submission date, or when a dissertation includes material already published in the literature.

A related point concerns the establishment of, or development of, any research collaboration. Given the centrality of publications as research outputs, and the importance of authorship as a recognition of research credit and accountability, it is sensible to ensure that expectations are clear when initiating a collaboration. Given that there may be somewhat different norms regarding authorship practices in different disciplines, this may be especially important in interdisciplinary work. No one should ever be surprised when their collaborators do, or do not, consider them to be authors of outputs. There should also be a shared plan of what outputs there will be, who is contributing what, and which contributions amount to authorship - ideally indicating expected ordering of authors on particular outputs. Such plans will often need to change as a project proceeds, but should do so by agreement. It is sensible to keep clear records of contributions (open to team members) to limit the likelihood of disputes about this later.

Some contributions will fall short of authorship, and may only deserve acknowledgement. However, in some studies there may be choices to be made *at the outset* over whether junior researchers (which may include graduate students) are to be considered as technicians (to follow instructions, and have their contribution acknowledged), or to be deliberately included in discussions of processes of conceptualization, design, and interpretation, at a level which will justify their status as authors. Best practice would seem to be to - whenever viable - look to engage junior colleagues at a level that justifies authorship. This is not only valuable for their publications list, but also because it provides research development for them and capacity building for the institution and wider field.

## References


1. Merton, R. K. Priorities in Scientific Discovery: A Chapter in the Sociology of Science. *Am. Sociol. Rev.* **1957**, *22*, 635–659.
2. Taber, K. S. *Classroom-based Research and Evidence-based Practice: An Introduction*, 2nd ed; Sage: London, 2013.
3. Phillips, D. C.; Burbules, N. C. *Postpositivism and Educational Research*; Rowman & Littlefield: Oxford, U.K., 2000.
4. Taber, K. S. The Natures of Scientific Thinking: Creativity as the Handmaiden to Logic in the Development of Public and Personal Knowledge. In *Advances in the Nature of Science Research - Concepts and Methodologies*; Khine, M. S., Ed.; Springer: Dordrecht, 2011; pp 51−74.
5. Box, G. E. P.; Draper, N. R. *Empirical Model-Building and Response Surfaces*; John Wiley: New York, 1987.
6. Taber, K. S. *Progressing Science Education: Constructing the Scientific Research Programme into the Contingent Nature of Learning Science*; Springer: Dordrecht, NL, 2009.
7. Lakatos, I. Falsification and the Methodology of Scientific Research Programmes. In *Criticism and the Growth of Knowledge*; Lakatos, I., Musgrave, A., Eds.; Cambridge University Press: Cambridge, 1970; Vol. 4, pp 91−196.
8. Taber, K. S. Who Counts as an Author When Reporting Educational Research? *Chem. Educ. Res. Pract.* **2013**, *14*, 5–8.
9. Gøtzsche, P. C.; Kassirer, J. P.; Woolley, K. L.; Wager, E.; Jacobs, A.; Gertel, A.; Hamilton, C. What Should Be Done To Tackle Ghostwriting in the Medical Literature? *PLoS Med.* **2009**, *6*, e1000023.
10. Annunziata, S.; Giordano, A. Authorship Problems in Scientific Literature and in Nuclear Medicine: the Point of View of the Young Researcher. *Eur. J. Nucl. Med. Mol. Imaging.* **2014**, *41*, 1251−1254.
11. Student, The Probable Error of a Mean. *Biometrika* **1908**, *6*, 1–25.
12. Susser, M. Authors and Authorship-Reform or Abolition? *Am. J. Public Health* **1997**, *87*, 1091−1092.
13. Kuhn, T. S. The Essential Tension: Tradition and Innovation in Scientific Research. In *The Essential Tension: Selected Studies in Scientific Tradition and Change*; Kuhn, T. S., Ed.; University of Chicago Press: Chicago, IL, 1959/1977; pp 225−239.
14. Begg, C. B.; Berlin, J. A. Publication Bias: A Problem in Interpreting Medical Data. *J. R. Stat. Soc. Series A (Statistics in Society).* **1988**, *151*, 419–463.
15. Taber, K. S. The End of Academic Standards? A Lament on the Erosion of Scholarly Values in the Post-Truth World. *Chem. Educ. Res. Pract.* **2018**, *19*, 9–14.
16. Taber, K. S. Methodological Issues in Science Education Research: a Perspective From the Philosophy of Science. In *International Handbook of Research in History, Philosophy and Science Teaching*; Matthews, M. R., Ed.; Springer Netherlands: Dordrecht, 2014; pp 1839−1893.

17. Rond, M. D.; Miller, A. N. Publish or Perish: Bane or Boon of Academic Life? *J. Manage. Inq.* **2005**, *14*, 321–329.
18. Taber, K. S. Non-random Thoughts About Research. *Chem. Educ. Res. Pract* **2013**, *14*, 359–362.
19. Moser, M. Concerns About Authorship and Bias in Scientific Publications. *J. Clin. Hypertens.* **2006**, *8*, 613–614.
20. Inge, M. T. Collaboration and Concepts of Authorship. *PMLA.* **2001**, *116*, 623–630.
21. Taber, K. S. *Modelling Learners and Learning in Science Education: Developing Representations of Concepts, Conceptual Structure and Conceptual Change to Inform Teaching and Research*; Springer: Dordrecht, NL, 2013.
22. Luria, A. R. *The Mind of a Mnemonist: A Little Book About a Vast Memory*; Harvard University Press: Cambridge, MA, 1987.
23. Peterson, R. F.; Treagust, D. F.; Garnett, P. Development and Application of a Diagnostic Instrument to Evaluate Grade-11 and -12 Students' Concepts of Covalent Bonding and Structure Following a Course of Instruction. *J. Res. Sci. Teach.* **1989**, *26*, 301–314.
24. Peterson, R. F.; Treagust, D. F.; Garnett, P. Identification of Secondary Students' Misconceptions of Covalent Bonding and Structure Concepts Using a Diagnostic Instrument. *Res. Sci. Educ.* **1986**, *16*, 40–48.
25. Taber, K. S; Tan, K. C. D. The Insidious Nature of 'Hard Core' Alternative Conceptions: Implications for the Constructivist Research Programme of Patterns in High School Students' and Pre-Service Teachers' Thinking About Ionisation Energy. *Int. J. Sci. Educ.* **2011**, *33*, 259–297.
26. Tan, K.-C. D.; Taber, K. S.; Goh, N.-K.; Chia, L.-S. The Ionisation Energy Diagnostic Instrument: a Two-Tier Multiple Choice Instrument to Determine High School Students' Understanding of Ionisation Energy. *Chem. Educ. Res. Pract.* **2005**, *6*, 180–197.
27. Tan, K.-C. D.; Taber, K. S. Ionization Energy: Implications of Pre-service Teachers' Conceptions. *J. Chem. Educ.* **2009**, *86*, 623–629.
28. British Educational Research Association. *Good Practice in Educational Research Writing*; British Educational Research Association: Southwell, Notts., U.K., 2000; p 7.
29. Kuhn, T. S. *The Structure of Scientific Revolutions*, 3rd ed.; University of Chicago Press: Chicago, IL, 1996.
30. Knorr Cetina, K. *Epistemic Cultures: How the Sciences Make Knowledge*; Harvard University Press: Cambridge, MA, 1999.
31. Aad, G.; et al. (ATLAS Collaboration). Observation of a New Particle in the Search for the Standard Model Higgs Boson With the ATLAS Detector at the LHC. *Phys. Lett. B* **2012**, *716*, 1–29.
32. Taber, K. S.; Brock, R.; Martínez Sainz, G. Thinking together, learning together, writing together: synergies and challenges in the collaborative supervisory relationship. In *Working Papers Series*; University of Cambridge Faculty of Education: Cambridge, 2016, pp 1–32.

33. King, J. J. T. How Many Neurosurgeons Does It Take to Write a Research Article? Authorship Proliferation in Neurosurgical Research. *Neurosurgery.* **2000**, *47*, 435–440.
34. Modi, P.; Hassan, A.; Teng, C. J.; Chitwood, W. R. How Many Cardiac Surgeons Does it Take to Write a Research Article?: Seventy Years of Authorship Proliferation and Internationalization in the Cardiothoracic Surgical Literature. *J. Thorac. Cardiovasc. Surg.* **2008**, *136*, 4–6.
35. Rahman, L.; Muirhead-Allwood., S. K. How Many Orthopedic Surgeons Does It Take to Write a Research Article? 50 Years of Authorship Proliferation in and Internationalization of the Orthopedic Surgery Literature. *Orthopedics* **2010**, *33*, 478.
36. Matzinger, P.; Mirkwood, G. In a Fully H-2 Incompatible Chimera, T Cells of Donor Origin Can Respond to Minor Histocompatibility Antigens in Association with Either Donor or Host H-2 Type. *J. Exp. Med.* **1978**, *148*, 84–92.
37. Hetherington, J. H.; Willard, F. D. C. Two-, Three-, and Four-Atom Exchange Effects in bcc 3He. *Phys. Rev. Lett.* **1975**, *35*, 1442–1444.
38. Alpher, R. A.; Bethe, H.; Gamow, G. The Origin of the Chemical Elements. *Phys. Rev.* **1948**, *73*, 803–804.
39. Zeller, B. On Translation and Authorship. *Meta* **2000**, *45*, 134–139.
40. Limerick, B.; Burgess-Limerick, T.; Grace, M. The Politics of Interviewing: Power Relations and Accepting the Gift. *Int. J. Qual. Stud. Educ.* **1996**, *9*, 449–460.
41. Intellectual Property Office. *Ownership of copyright works*; 2014; available from https://www.gov.uk/guidance/ownership-of-copyright-works#works-created-by-students (accessed March 17, 2018).
42. Taber, K. S. Ethical Considerations of Chemistry Education Research Involving "Human Subjects". *Chem. Educ. Res. Pract.* **2014**, *15*, 109–113.
43. Braz, A. Collaborative Authorship and Indigenous Literatures. *CLCWeb: Comparative Literature and Culture* **2011**, *13*Article 5.
44. Dumais, S. A. Cultural Capital, Gender, and School Success: The Role of Habitus. *Sociol. Educ.* **2002**, *75*, 44–68.
45. Taber, K. S; Student, T. A. How Was It For You?: the Dialogue Between Researcher and Colearner. *Westminster Studies in Education.* **2003**, *26*, 33–44.
46. Wislar, J. S.; Flanagin, A.; Fontanarosa, P. B.; DeAngelis, C. D. Honorary and Ghost Authorship in High Impact Biomedical Journals: a Cross Sectional Survey. *BMJ* **2011**, *343*, d6128.

## Chapter 2

# Case Study: An Editor's Perspective on Authorship

**Jeffrey Kovac**

**Department of Chemistry, University of Tennessee, Knoxville, Tennessee 37996, United States**
***E-mail: jkovac@utk.edu.**

The following case study explores an ethical issue that the editor of a journal might encounter: some of the authors of a multi-author article that has been submitted for publication raise concerns regarding the final version that is to be sent out for review. The editor must try to resolve this authorship dispute. Although the ethical issues are quite accessible, this case might not be appropriate for students who are less familiar with the details of the publication process.

## An Editor's Dilemma

Tim Amis, the editor of a major journal in materials chemistry, was trying to identify appropriate reviewers for an article that had just been submitted. The research reported was a collaboration among three research groups at different universities. The corresponding author was Rebecca Smith, a chaired professor at Big State University. As Amis was preparing to send messages to three potential reviewers, he received an e-mail message from one of the other authors, Michael Gross, a senior professor at Famous Private University. In that message Gross explained that members of his research group were major contributors to the article submitted by Smith; they felt that the experiments were both correct and important, but they had some serious concerns about the final version of the paper. "We feel that Smith overinterpreted the results and drew at least one unwarranted conclusion. We expressed our concerns as the article was being written, but Smith ignored them and submitted anyway. I don't feel comfortable with the conclusion as written. I think this needs to be resolved before the paper is sent out for review."

What should Tim Amis do? Here are several possibilities, but you may think of others.

1. Should he send the article to the reviewers in its present form hoping that the reviews will provide a way to resolve this issue?
2. Should he write to Robert Smith and tell him that the authors need to resolve their differences before proceeding further?
3. Should he ask Michael Gross to send him a detailed explanation of his objections and then write to Smith, providing suggestions for potential revisions of the article that will make Gross more comfortable with the conclusions?
4. Does he need to contract the third research group to find out if they agree with the submitted version? If they agree with the current version, what should he do? If they have the same or different reservations, how should he proceed?
5. Should Amis contact Smith's department with concerns about his conduct? On the face of it, Smith has violated norms of professional ethics by submitting an article that was not approved by all the authors.

Chapter 3

# Contributorship and Authorship Hierarchy as a Form of Credit

**Cory Craig***

**Physical Sciences and Engineering Library,**
**University of California Davis, 1 Shields Avenue,**
**Davis, California 95616, United States**
***E-mail: cjcraig@ucdavis.edu.**

Authorship is central to academic scholarship and reward. It provides both credit and accountability and plays a critical role in promotion and tenure for academic scientists. The increasing numbers of authors on scientific papers are stretching, challenging, and changing notions of authorship and raising issues of credit and accountability. Because conventions for listing author names vary, determining who contributed what to a research publication can be next to impossible. Many are suggesting that contributions to published research should be made transparent and evident to readers. This chapter describes several substantive efforts focused on making author contribution statements transparent and summarizes criteria which are key to including author contributions in the scholarly record.

## Authorship in Academic Scholarship

Authorship is central to academic scholarship and reward. It provides both recognition, or credit, and accountability; and it plays a critical role in promotion, tenure, and funding opportunities for academic scientists (*1–3*). Prior to the 1900s, single or co-authorship was the norm, but over the last 50 to 100 years, authorship in science has changed, showing an increase in the number of authors on scientific papers.

From the 1930s to the 1960s, the average number of authors on scientific papers was approximately two (*4*, *5*). By 2000, the average number of authors in high-ranked biomedical journals was seven, and the maximum number of authors in MEDLINE articles was 38 (*4, 6*). Analyses of articles indexed by Web of Science have shown that the maximum number of authors on a paper increased from 553 (in 1992) to 3,179 (in 2011); from 1998 to 2011, papers with >50 authors, >100 authors, >200 authors, >500 authors, >1000 authors, all increased in number; and, from 2002 to 2011, the number of physical sciences and biological sciences papers with over 100 authors showed significant increases (*7–9*). In May 2015, a well-publicized high-energy physics paper with 5,154 authors was described as breaking the record for the highest number of authors for a journal article (*10*).

Several driving forces are contributing to this growth in number of authors per paper. Authorship, at least in science, is often a collective activity that relies on complex teams of researchers, representing multiple institutions and crossing disciplinary boundaries. In addition, a relatively small amount of federal research funds are awarded to individual investigators, as opposed to teams (*11*). Other factors include academic reward systems, increasing interdisciplinary and team science initiatives, and the ease of Internet collaboration (*4*).

## Ethical Considerations

Multi-authored works are stretching, challenging, and changing notions of authorship and raising issues of credit and accountability. Standard author disputes, often managed by publishers, can occur in any research group. These typically focus on authority, inclusion, and ordering of author's names. But large multi-author publications have raised additional questions, including whether or not all authors: (1) are able to support/defend claims of the paper; and (2) should be equally credited and accountable for all claims of the paper.

An analysis of data on contributorship from the journal *PLOS ONE*, found that labor was distributed in most disciplines, and often the person writing the article was not the one who did the experiments (*12*). Lariviere (*12*) has pointed out that collective authorship, lacking clear contribution statements, has a "double consequence"; because each author receives publication credit, recognition is increased but accountability is divided among a larger number of authors. Yet, collaboration allows important and useful contributions in science; these contributions should be encouraged and rewarded, but the current system of authorship does not do that (*13*). Making author contributions evident can greatly reduce ethical dilemmas, reduce author disputes, and make science more transparent.

## Authorship Definitions

Despite the importance of authorship to academic scholarship, no universal criteria exist for conferring authorship status (*1, 2, 11*). Publishers and academic organizations within different disciplines have developed criteria defining what counts as authorship. Selected examples are given below.

### American Chemical Society (ACS)

Authors are defined as: "all those persons who have made significant scientific contributions to the work reported and who share responsibility and accountability for the results" (*14, 15*). Responsibilities of authors include:

- appropriately recognizing the contributions of technical staff and data professionals;
- including all appropriate persons as co-authors (and none that are inappropriate);
- obtaining each author's consent to be a co-author (*14, 15*).

### International Committee of Medical Journal Editors (ICMJE)

The ICMJE gives four criteria and recommends that authors meet all of them:

- Substantial contributions to the conception or design of the work; or the acquisition, analysis, or interpretation of data for the work;
- Drafting the work or revising it critically for important intellectual content;
- Final approval of the version to be published;
- Agreement to be accountable for all aspects of the work in ensuring that questions related to the accuracy or integrity of any part of the work are appropriately investigated and resolved (*16*).

### Council of Science Editors (CSE)

"Authors are individuals identified by the research group to have made substantial contributions to the reported work and agree to be accountable for these contributions. In addition to being accountable for the parts of the work he or she has done, an author should be able to identify which of their coauthors are responsible for specific other parts of the work. In addition, an author should have confidence in the integrity of the contributions of their co-authors. All authors should review and approve the final manuscript" (*17*).

### American Physical Society (APS)

"Authorship should be limited to those who have made a significant contribution to the concept, design, execution or interpretation of the research study. All those who have made significant contributions should be offered the opportunity to be listed as authors. Other individuals who have contributed to the study should be acknowledged, but not identified as authors" (*18*).

All these definitions agree that authorship requires significant or substantial contributions to the work being reported. Additional central themes include: taking responsibility or accountability for contributions; appropriately recognizing all contributions to the work; and obtaining consent of all co-authors (to be authors and to be accountable). A problem that emerges is that the roles and types of contributions are not defined. Requirements for authorship can also be specific to a researcher's lab, and these may even define the types of contributions required (*19*). But even for comprehensive, well-thought out rubrics, contributions of authors may not be transparent in the final publication, depending on the practices of the journal or publisher.

## Authorship Hierarchy

A related issue is the order of authors listed, or authorship hierarchy. Practices vary between and within disciplines. Author names can be listed according to a wide range of methods, including: alphabetical, weighted, reverse seniority, or the "sandwich model", where the first author does most of the work, the senior author is listed last, and everyone else is in the middle (*2*, *13*, *15*).

Most readers feel the order of authors indicates something about who did what on the paper. A study focused on promotion committees at U.S. medical schools examined perceptions of author contributions based on relative author position and found that researchers do apportion credit by author position (*3*), but, because conventions vary and are typically understood only by experienced readers, it may not be possible to decode an author list to determine who contributed what to the final paper (*20*). This task is even more difficult for papers with large numbers of authors.

A systematic review of authorship practices found that for most science researchers, the preferred method for determining authorship order is the amount of work done, not prestige or status (*2*). Lacking an explicit statement of contributorship, readers, including faculty promotion committees, are likely to implicitly allocate authorship credit without any defined standard (*3*, *21*). Given the importance of authorship within academic scholarship and the importance of research findings for society, many are suggesting that contributions to authorship should be transparent, accurate, and evident to readers (*3*, *4*, *20*, *22–25*).

## Contributor Statements: A Step Towards Transparency

An examination of authorship policies for a random sample of 600 journals from the Journal Citation Reports database found 62.5% of these journals had an authorship policy, but only 5.3% have a policy requiring that authors describe their contributions (*26*). For publishers that require or invite contribution statements, there can be great variation in how this information is obtained; some provide a predefined list of roles, others allow free-text statements from authors, and some publishers collect contributorship information but do not actually publish it (*4*, *25*). Several substantive efforts focused on providing author contribution statements have emerged. This section will describe several of these efforts and discuss next steps.

**Project CRediT (Contributor Roles Taxonomy)**

Project CRediT (Contributor Roles Taxonomy) is a high-level classification of the diverse roles that contribute to published research output in the sciences. It was developed to provide transparency to all contributions to scholarly published work and to improve systems of credit, attribution, and accountability (*4*, *27*). Table 1 provides a listing and definitions of the contributor roles that make up the CRediT Taxonomy.

**Table 1. Project CRediT: Contributor Roles Taxonomy. Adapted with permission from ref. (*27*). Copyright 2006-2018 CASRAI.**

| *Contributor Role* | *Definition* |
|---|---|
| Conceptualization | Ideas; formulation or evolution of overarching research goals and aims |
| Data curation | Management activities to annotate (produce metadata), scrub data, and maintain research data (including software code, where it is necessary for interpreting the data itself) for initial use and later reuse |
| Formal Analysis | Application of statistical, mathematical, computational, or other formal techniques to analyze or synthesize study data |
| Funding acquisition | Acquisition of the financial support for the project leading to this publication. |
| Investigation | Conducting a research and investigation process, specifically performing the experiments, or data/evidence collection |
| Methodology | Development or design of methodology; creation of models |
| Project Administration | Management and coordination responsibility for the research activity planning and execution |
| Resources | Provision of study materials, reagents, materials, patients, laboratory samples, animals, instrumentation, computing resources, or other analysis tools |
| Software | Programming; software development; designing computer programs; implementation of the computer code and supporting algorithms; testing of existing code components |
| Supervision | Oversight and leadership responsibility for the research activity planning and execution, including mentorship external to the core team |
| Validation | Verification, whether as a part of the activity or separate, of the overall replication or reproducibility of results, experiments, and other research outputs |

*Continued on next page.*

**Table 1. (Continued). Project CRediT: Contributor Roles Taxonomy.**

| *Contributor Role* | *Definition* |
|---|---|
| Visualization | Preparation, creation and/or presentation of the published work, specifically visualization and data presentation |
| Writing – Original Draft | Preparation, creation and/or presentation of the published work, specifically writing the initial draft (including substantive translation) |
| Writing – Review & Editing | Preparation, creation and/or presentation of the published work by those from the original research group, specifically critical review, commentary, or revision, including pre- or postpublication stages |

The fourteen roles identified in the CRediT taxonomy include, but are not limited to, traditional authorship roles. The roles given are not intended to define what constitutes authorship, but instead to capture all the work that allows scholarly publications to be produced. Recommendations for applying the CRediT taxonomy are:

- **List All Contributions:** all contributions should be listed, whether they are from those formally listed as authors or individuals named in acknowledgements;
- **Multiple Roles Possible:** individual contributors can be assigned multiple roles, and a given role can be assigned to multiple contributors;
- **Degree of Contribution Optional:** where multiple individuals serve in the same role, the degree of contribution can optionally be specified as 'lead', 'equal', or 'supporting';
- **Shared Responsibility:** corresponding authors should assume responsibility for role assignment, and all contributors should be given the opportunity to review and confirm assigned roles;
- **Make CRediT Machine Readable:** CRediT tagged contributions should be coded in JATS XML v1.2 (*4*, *27*, *28*).

The CRediT Taxonomy is now used by over 100 journals. Early adopters include Cell Press journals, PLOS journals, eLife, and GigaScience, as well as Aries Systems, a manuscript submission system used by many journal publishers (*29*). Table 2 lists publishers, integrators and publishing outlets that currently use the CRediT Taxonomy; the most current listing will be available via the Project CRediT webpage (*29*). It should be noted that current implementations of CRediT vary widely. Some publishers make it available, but do not require its use; other publishers collect contribution data but do not make it publically available; and, importantly, few publishers are currently making contribution data machine readable in the article XML. The CRediT Taxonomy is also working toward integration with ORCID (*30*). The aim is to produce a taxonomy that is simple to use but capable of representing the wide range of contributions to published output in science (*25*, *27*). Additional benefits the CRediT Taxonomy may provide include making it easier to identify both potential collaborators and candidates for peer review and reducing the number of author disputes which journal editors must manage (*4*, *27*).

**Table 2. Publishers, Integrators, and Publishing Outlets Using CRediT**

| *Publishers* | | | *Integrators & Publishing Outlets* |
|---|---|---|---|
| American Association of Petroleum Geologists | F1000 Research | Oxford University Press | Allen Press/ Peer Track |
| BMJ Open Science | Geological Society of London | Public Library of Science | Aries Systems/ Editorial Manager |
| British Psychological Society | Health & Medical Publishing Group | SAE International | Coko Foundation/ xPub |
| Cell Press | International Centre of Insect Physiology and Ecology | SLACK Incorporated | HRB Open Research |
| Dartmouth Journal Services | The Journal of Bone & Joint Surgery | Springer | River Valley/ ReView |
| De Gruyter Open | KAMJE Press | Springer Publishing Company | Gates Open Research |
| Duke University Press | Lippincott Williams & Wilkins | Wiley VCH | Wellcome Open Research |
| eLife | MA Healthcare | Wolters Kluwer | |
| Elsevier | MIT Press | | |
| Evidence Based Communications | Oman Medical Specialty Board | | |

The CRediT Taxonomy, along with guidelines for using it, is hosted by CASRAI (Consortia Advancing Standards in Research Administration Information), which works to improve information flow within and between research stakeholders (*25*, *29*). Project CRediT working groups are now focusing on encouraging and supporting implementations of the taxonomy; evaluating further development of the taxonomy; and increasing buy-in by publishers, researchers, and other stakeholders.

### FORCE11 Attribution Working Group

FORCE11 is a community of scholars, librarians, archivists, publishers and research funders working to advance scholarly communication through the effective use of information technology. The FORCE11 Attribution Working Group was formed out of the FORCE2015 "Contribution and Attribution in the Context of the Scholar" workshop. The group is working to:

- Collate and review existing efforts on scholarly contribution taxonomies;
- Determine if a consensus implementation, which would meet the requirements for all projects and would include the capability to be extensible from a core taxonomy, is possible (*25*, *31*).

### OpenVIVO Contribution Role Ontology (CRO)

OpenVIVO is a free, open-hosted semantic web platform that gathers and shares open data about scholarship. The VIVO platform provides access to data about the scholarly work of its participants. It is based on the VIVO open source platform and membership is open to anyone who creates an OpenVIVO profile (*32*). OpenVIVO is interoperable with ORCID.

OpenVIVO uses classes, data properties, and object properties from twenty different ontologies to represent the scholarship of its participants (*32*). VIVO relies on the basic structure of the semantic triple (a set of three entities and a description of the relationships between them) to record all information. The OpenVIVO system includes a Contribution Role Ontology (CRO), a model of attribution for scholarly output which provides 60 contribution roles (*32*). Contributor roles are based on:

- the CRediT Taxonomy;
- author roles collected at the FORCE 2016 Conference, and the FORCE11 Attribution Working Group; and
- Library of Congress cataloging standards (MARC code of relator terms) that designate the agent (i.e., the individual or entity responsible for creating a work, or author) to bibliographic resource relationship (*32*).

To indicate contributions to scholarly works in the OpenVIVO system, researchers create an OpenVIVO profile, and then complete the claim process in OpenVIVO. OpenVIVO retrieves metadata for a given work, and the researcher is asked to identify their contributions. See Figure 1 for an image of the OpenVIVO claim process and a partial listing of roles from the Contribution Role Ontology. For full listing of contributor roles from the OpenVIVO CRO, see Ilik et al. (*32*) or the project's OpenRIF GitHub repository: https://github.com/openrif/contribution-ontology/tree/reorganized-hierarchy.

*Figure 1. Adapted from "OpenVIVO Contribution Role Hierarchy" by Ilik, V.; Conlon, M.; Triggs, G.; White, M.; Javed, M.; Brush, M.;Gutzman, K.; Essaid, S.; Friedman, P.; Porter, S.; Szomszor, M.; Haendel, M. A.; Eichmann, D.; Holmes, K. L. Frontiers in Research Metrics and Analytics **2018**, 2, 1-11, licensed under CC BY.*

Key differences between the OpenVIVO CRO and the CRediT Taxonomy include:

- **Number of contributor roles:** The OpenVIVO CRO lists 60 contributor roles, and may cover a wider range of research than the 14 roles given in CRediT Taxonomy, yet the larger number of roles may be more difficult to standardize.
- **When and how contribution data are assigned:** Because the CRediT Taxonomy is implemented by publishers, contribution data are collected in a single step for all authors as part of the article submission process (it is recommended that all authors review and agree on author contributions); with the OpenVIVO CRO, each author identifies their own contributions within their individual OpenVIVO profile.
- **Where data reside**: CRediT Taxonomy recommendations indicate that contribution data should be machine readable and part of the article XML metadata to ensure that these data are discoverable through data mining and online searching. Currently only a few publishers are known to be implementing this recommendation. With OpenVIVO, because contribution data are recorded in the OpenVIVO platform, all contribution data entered are machine readable. In addition, because

contribution data are assigned in the OpenVIVO system, it does not require buy-in or implementation by publishers.

All three efforts have a number of important features in common as well. The CRediT Taxonomy, the FORCE11 Attribution Working Group, and the OpenVIVO CRO all: (1) use and value similar methods: working with the scholarly community to identify, test, and refine contributor roles; (2) recognize the importance of making these data transparent; and (3) share a common goal to create an intuitive, easy to use taxonomy that makes author contributions openly available and easily electronically accessible.

## Concluding Remarks

Long-used bibliographic conventions for describing authorship have simply not kept pace with the semantic capabilities of web publishing; publishers are indicating that it is now technically feasible to provide an additional information layer that standardizes and identifies the contributions of authors (*25*). The CRediT Taxonomy and the OpenVIVO CRO represent significant progress to efforts that have been going on for 20 years (*25*, *33*). With technical advancements and these taxonomies, there is growing interest from researchers, publishers, academic institutions, and funding agencies in making author contributions to published research transparent and accessible to both readers and those who evaluate the work of academic scientists (*25*).

### Recognizing Author Contributions: Criteria for Success

The U.K. Academy of Medical Sciences has examined how collaboration and team science can be recognized and rewarded in academia and has issued an extensive and well-thought-out report. Two of the ten recommendations provided address making author contributor information transparent:

Recommendation 1: "All research outputs and grants should include open, transparent, standardized and structured contribution information" (*25*).

Recommendation 2: "The most effective way of providing contribution information will be an open and transparent research information infrastructure which links all research inputs and outputs to individual contributors" (*25*).

Recommendation 1 also indicates that publishers and the research community should work with Project CRediT and related initiatives, to create a standardized author contribution framework (*25*).

The report states that such a system is essential and also identifies key criteria for a successful framework:

- Standardized categories of contribution, so that individuals encounter the same system in all work, whether as a researcher or an appraiser.
- Guidelines on how to allocate contributions in a transparent and consensually agreed manner.

- The allocation of contributions needs to be agreed by all listed on the publication upon submission to ensure accuracy and fairness, and to disincentivize the appearance of guest or honorary authors. Any changes will also need to be agreed by all those listed. Alongside the information being 'authorized' by the publisher, this will be necessary to provide appraisers with confidence of the information's accuracy.
- The system must link to other research information systems by capturing individual digital identifiers for everyone listed on the publication.
- The system must be as simple as possible, whilst remaining fit for the purposes of researchers, employers and funders. This was particularly important for funders and employers.
- The contribution information must be prominent and easily electronically accessible.
- Our researchers' workshop suggested a change from the current format, such as a grid or heatmap layout as a way of visually summarizing an individual's contributions (*25*).

An ideal system would be interoperable with both ORCID (linking to researchers' ORCID iDs), as well as digital object identifiers (DOIs) for journal articles (*25*). Existing taxonomies might also look to further refinements. A single taxonomy offers obvious benefits to standardization but may not be comprehensive enough for all science disciplines. In addition, descriptors that communicate who did the bulk of the work on a project might also be useful. These could be either quantitative (percent contribution) or qualitative descriptors (text to identify 'principal' or 'major' contributors) (*25*).

Given the importance of authorship and author contributions to academic scholarship and reward, it is time to move beyond the author paradigm and find ways to represent and acknowledge all contributions that create scholarly publications. This will require continued involvement from researchers, publishers, and all stakeholders in academic scholarship.

## Acknowledgments

Two anonymous reviewers provided very useful comments and insights that resulted in improvements to the chapter. Simon Kerridge reviewed the chapter, and provided very useful feedback.

## References

1. *What Constitutes Authorship? COPE Discussion Document*; COPE Committee on Publication Ethics 2014.
2. Marusic, A.; Bosnjak, L.; Jeroncic, A. A Systematic Review of Research on the Meaning, Ethics and Practices of Authorship across Scholarly Disciplines. *PLoS One* **2011**, *6*, e23477.

3. Wren, J. D.; Kozak, K. Z.; Johnson, K. R.; Deakyne, S. J.; Schilling, L. M.; Dellavalle, R. P. The Write Position. A Survey of Perceived Contributions to Papers Based on Byline Position and Number of Authors. *EMBO Rep.* **2007**, *8*, 988–91.
4. Brand, A.; Allen, L.; Altman, M.; Hlava, M.; Scott, J. Beyond Authorship: Attribution, Contribution, Collaboration, and Credit. *Learn Publ.* **2015**, *28*, 151–155.
5. Clarke, B. L. Multiple Authorship Trends in Scientific Papers. *Science* **1964**, *143*, 822–824.
6. *Number of Authors per MEDLINE®/PubMed® Citation.* https://www.nlm.nih.gov/bsd/authors1.html (accessed October 15, 2017).
7. King, C. Crowd Control? Multiauthor Papers Appear to Level Off in Recent Years. *ScienceWatch Newsletter.* **2004**July-August.
8. King, C. Multiauthor Papers Redux: a New Peek at New Peaks. *ScienceWatch Newsletter.* **2007**November-December.
9. King, C. Multiauthor Papers: Onward and Upward. *ScienceWatch Newsletter.* **2012**July.
10. Castelvecchi, D. Physics Paper Sets Record with More than 5,000 Authors. *Nature* **2015**May15. https://www.nature.com/news/physics-paper-sets-record-with-more-than-5-000-authors-1.17567 (accessed May 15, 2018).
11. Cronin, B. Hyperauthorship: A Postmodern Perversion or Evidence of a Structural Shift in Scholarly Communication Practices. *J. Am. Soc. Inf. Sci. Technol.* **2001**, *52*, 558–569.
12. Lariviere, V.; Desrochers, N.; Macaluso, B.; Mongeon, P.; Paul-Hus, A.; Sugimoto, C. R. Contributorship and Division of Labor in Knowledge Production. *Soc. Stud. Sci.* **2016**, *46*, 417–435.
13. Gymrek, M. Middle Author Dilemma: How to Recognize Critical Contributions of Multidisciplinary Teams. In *8th Conference on Open Access Scholarly Publishing (COASP)* Arlington, Virginia, U.S.A., 2016.
14. *Ethical Guidelines to Publication of Chemical Research.* https://pubs.acs.org/userimages/ContentEditor/1218054468605/ethics.pdf (accessed January 1, 2018).
15. Hammes, G. G. Ethics in Scientific Publication. In *The ACS Style Guide*; Coghill, A. M., Garson, L. R., Eds.; American Chemical Society: Washington, DC, 2006; pp 3−16.
16. *Defining the Role of Authors and Contributors*; International Committee of Medical Journal Editors (ICMJE): http://www.icmje.org/recommendations/browse/roles-and-responsibilities/defining-the-role-of-authors-and-contributors.html (accessed May 15, 2018).
17. *White Paper on Publication Ethics: 2.2 Authorship and Authorship Responsibilities*; Council of Science Editors, 2012. https://www.councilscienceeditors.org/resource-library/editorial-policies/white-paper-on-publication-ethics/2-2-authorship-and-authorship-responsibilities/ (accessed May 15, 2018).
18. *APS Guidelines for Professional Conduct.* https://www.aps.org/policy/statements/02_2.cfm (accessed January 1, 2018).

19. Kosslyn, S. M. *Criteria for Authorship*. https://kosslynlab.fas.harvard.edu/files/kosslynlab/files/authorship_criteria_nov02.pdf (accessed October 15, 2017).
20. Frische, S. It is Time for Full Disclosure of Author Contributions. *Nature* **2012**, *489*, 475–475.
21. Tarnow, E.; De Young, B. R.; Cohen, M. B. Coauthorship in Pathology, a Comparison with Physics and a Survery-Generated and Member-Preferred Authorship Guideline. *MedGenMed.* **2004**, *6*, 1–2.
22. Allen, L.; Brand, A.; Scott, J.; Altman, M.; Hlava, M. Credit Where Credit is Due. *Nature* **2014**, *508*, 312–313.
23. Pearson, H. Credit Where Credit's Due. *Nature* **2006**, *440*, 591–592.
24. Canadian Academy of Health Sciences. *Academic Recognition of Team Science: How to Optimize the Canadian Academic System*; The Expert Panel on Academic Recognition of Team Science in Canada, CAHS, Ottawa (ON), 2017.
25. *Improving Recognition of Team Science Contributions in Biomedical Research Careers*; Academy of Medical Sciences: London, UK, 2016. https://acmedsci.ac.uk/policy/policy-projects/team-science (accessed May 15, 2018).
26. Resnik, D. B.; Tyler, A. M.; Black, J. R.; Kissling, G. Authorship Policies of Scientific Journals. *J Med Ethics.* **2016**, *42*, 199–202.
27. *CRediT: Contributor Role Taxonomy. Going Beyond Authorship.* https://docs.google.com/document/d/1aJxrQXYHW5U6By3KEAHrx1Iho6ioeh3ohNsRMwsoGPM/edit (accessed October 15, 2017).
28. *Project CRediT Contributor Roles.* http://dictionary.casrai.org/Contributor_Roles (accessed October 15, 2017).
29. *Project CRediT.* http://docs.casrai.org/CRediT (accessed October 15, 2017).
30. Haak, L. L. *Contributor Role Pilot, Project CRediT (slides)* https://www.slideshare.net/CASRAI/20141210-haak-project-credit-1-42704799 (accessed January 1, 2018).
31. *Force11 Attribution Working Group.* https://www.force11.org/group/attributionwg (accessed January 15, 2018).
32. Ilik, V.; Conlon, M.; Triggs, G.; White, M.; Javed, M.; Brush, M.; Gutzman, K.; Essaid, S.; Friedman, P.; Porter, S.; Szomszor, M.; Haendel, M. A.; Eichmann, D.; Holmes, K. L. OpenVIVO: Transparency in Scholarship. *Front Res Metr Anal.* **2018**, *2*, 1–11.
33. Smith, R. Let's Simply Scrap Authorship and Move to Contributorship. *BMJ (Clinical research ed.)* **2012**, *344*, e157.

## Chapter 4

# Case Study: Contributorship and Authorship Hierarchy as a Form of Credit

**John D'Angelo**[*]

**Division of Chemistry, Alfred University, Alfred, New York 14802, United States**
**[*]E-mail: dangelo@alfred.edu.**

Contributorship and authorship hierarchy, often manifested as author order, can be a hotly contested issue. Herein are case studies both real and hypothetical that depict the sensitivities associated this hierarchy.

## Imagine This...

Your research group submits a publication to a high-profile journal, where it is accepted and eventually appears in print. Suddenly, however, one of the authors objects to the author order and even claims that the paper was published without their consent. The paper was similar (having a different author order and other significant edits, including new experiments) to a previously submitted but rejected paper; there were no objections to that submission. The retraction of this paper, which occurred because the dispute could not be resolved, will severely hamper the fellowship application you are about to finish.

*Questions to ponder:*

- Why wouldn't the approval for a previous submission carry over to an approval of a second submission?
- If the edits were significant enough, should the author order be impacted, and, if so, does the inclusion of new experiments reach this threshold?
- In cases where author order is **not** explicitly implicative of contribution, should author order be grounds for objecting to a publication?

Something very much like this has happened. At the core of the issue was the fact that one of the authors, Rodolfo Biekofsky, challenged the author order, specifically, his position as the eighth author. He went on to claim that, not only was this paper published without his consent, but that a similar paper submitted to four other journals listed him as a first author. The potential inappropriateness of the multiple submissions aside, one must wonder what prompted the change in author order, if Biekofsky was telling the truth (which is not immediately clear). Notably, his claim did not state they were *simultaneous* submissions. Biekofsky also provided evidence in the form of earlier purported drafts of initial versions of the paper that list him as first author and joint corresponding author. He claimed that he was never notified of the latest submission by his co-workers. The journal, in recognition of the potential impact of the work, gave the authors an opportunity to resolve the issue, but, when they failed to do so, the journal decided to retract the paper. Biekofsky went on to make other salacious claims against his co-authors that are unrelated to this publication, and the interested reader is encouraged to follow the link below for more information.

*Questions to ponder*

- What other options besides retraction may a journal employ in cases where the science of a paper is not in question? Could something like a note or addendum be published to describe the dispute?
- As a retraction doesn't erase the file from the computers of everyone who has viewed or downloaded it before it was retracted, what's the point of a retraction in a case like this? What, if anything, does it really serve?

Source:

Han, A.P. Retraction Watch. *Dispute over author order torpedoes paper on syndrome linked to autism.* October 11, 2017. http://retractionwatch.com/2017/10/11/dispute-author-order-torpedoes-paper-syndrome-linked-autism/#more-52077 (accessed January 22, 2018).

## Now Imagine This…

You are collaborating with another lab and finish writing a paper, on which you are the corresponding author. As the corresponding author, it is you who will complete the task of submitting the paper and detailing the roles of the authors. The work was done by one of your students and one student in the collaborative lab. On one of the final drafts, in addition to adding the experimental protocols the lab used, you notice that the collaborator has suddenly added two previously unmentioned authors. Given the regulations regarding authorship established by your field and the journal's policy of identifying the contributions of all the authors, you email the collaborator to determine these students' specific roles in the research and find the answers to be slightly dodgy, not directly answering your question. The correspondence is amicable but leaves you unsure of the next step.

*Questions to ponder:*

- Should you press your collaborator for more details?
- If you do press for more details, how should you do this? If not, without a firm answer from your collaborator, how should you list the contributions of the newly-added authors?

Such scenarios have actually occurred. In 2017, Mexican chemist Sylvain Bernès, while putting the finishing touches on a paper, contacted the collaborating lab, with whom he had collaborated for many years, asking the head of the lab to review the paper and add any additional authors worthy of co-authorship. What Bernès received in return concerned him: the collaborator added three authors without any justification or explanation. To find out how Bernès responded, consult the source below.

*Questions to ponder:*

- What, if anything, justifies Bernès being concerned?
- Should Bernès simply trust his collaborator and add the authors, since he did ask the collaborator to add any worthy authors?

## Source

Stern, V. Retraction Watch. *Unintended consequences: How authorship guidelines destroyed a relationship.* July 5, 2017. https://retractionwatch.com/2017/07/05/unintended-consequences-guidelines-destroyed-relationship/ (accessed January 22, 2018).

## Chapter 5

# You Stole My Invention! Authorship and Inventorship Considerations in Honoring Non-Disclosure Agreements

**Justin L. Krieger*** **and Akkad Y. Moussa**

**Kilpatrick Townsend & Stockton, LLP**
**607 14th Street NW, Suite 900**
**Washington, DC 20005, Unnited States**
***E-mail: jkrieger@kilpatricktownsend.com.**

The similarities and differences between authorship and inventorship are subtle, often leading to difficulties in deciding who should be designated an author on a work of authorship or an inventor on a patent application. In addition, obligations created by non-disclosure agreements (NDAs) can impact these important decisions as well as the timing of public disclosures and patent filings. This chapter identifies common problems and pitfalls in determining inventorship and authorship and provides guidance in assessing these vital decisions while honoring NDA obligations.

## Introduction

Non-disclosure agreements, or NDAs, have become a way of life among companies operating in research or technology, providing a valuable mechanism for protecting confidential corporate information. Although it may go by a different name—confidentiality agreement, proprietary information agreement, secrecy agreement, or even a component of an employment agreement—every NDA does the same thing: it creates legal obligations to keep information secret. Oftentimes, significant corporate value resides in confidential information, including trade secrets, rendering its potential disclosure a very real risk to corporations and their stockholders. NDAs create confidential relationships between two or more parties to protect the particular information specified, making them powerful legal tools for maintaining valuable corporate assets.

Whether it is a question of protecting a trade secret, limiting a disclosure in anticipation of a press release, or ensuring that an invention is kept secret before filing a patent application, a lot can ride on the specific terms of the NDA.

Competing with the desire to keep corporate information secret is the drive, in particular among the individuals performing the research, to publish articles and file patent applications concerning their discoveries. In many cases, the value in obtaining patents and the reputational benefits associated with publishing articles outweigh the value in keeping corporate information as a trade secret. Moreover, when research projects involve multiple individuals and/or companies, the different entities may have quite divergent views toward who should be named as an author or inventor. These conflicts can lead to important ethical and legal issues, such as: (1) how does one properly determine authorship and inventorship, and (2) how can NDA obligations impact one's ability to publish articles or file patent applications on his or her research?

## Overview of Authorship and Inventorship

### Defining Authorship

Determining authorship can be a difficult task involving both ethical and legal concerns. Academia, particularly in the sciences, has long been aware of the importance of self-imposed ethical standards. Without such standards, the public's trust in academic research and scholarship would rapidly dwindle, causing the usefulness and even the viability of academic research to suffer. Many institutions, from universities to publishers, have therefore developed and issued guidelines for researchers to ensure compliance. Included among these guidelines is the requirement that all authors be so identified in a publication. To facilitate that requirement, the guidelines typically include instructions for identifying and attributing authorship standards for academic papers. Although each institution's definition may vary slightly, a person is typically considered an author of a paper when he or she made substantial contributions to the work and somehow agreed to publication. Yale University, for example, defines an author as a contributor involved in planning, contributed some component of the work that led to the paper, drafted or revised the paper, and approved the final version of the paper (*1*). The International Committee of Medical Journal Editors (ICMJE) defines authorship using similar criteria and adds a fourth—an author must agree to be accountable for the accuracy and integrity of the work (*2*). The ACS itself imposes similar requirements, defining an author as any person who made "significant scientific contributions to the work reported and who share[s] responsibility and accountability for the results (*3*)".

According to these guidelines, however, it is not enough to merely attribute authorship credit to those who satisfy the specified criteria. A paper must also recognize those individuals who meaningfully contributed despite falling short of the requirements for authorship. That recognition is often made in a footnote or an "Acknowledgments" section of the paper. Although they may not have been as deeply involved in the research or drafting of the manuscript, those contributors still deserve credit. Each institution's guidelines may impose different requirements for how to go about giving that credit—the ICMJE, for example, requires that the specific contributions of non-author contributors be identified (*4*)—but ethical standards require that acknowledgments be made regardless of authorship status.

However, correctly identifying authors is not simply a matter of ethical compliance; it also impacts legal rights. Notions of authorship appear in a number of aspects of the law, but none more readily than in the law of copyrights, which provides legal protection over "original works of authorship fixed in any tangible medium (*5*)" Properly determining authorship is especially important in copyright law, because according to US law, unless there is a separate contractual obligation or the work constitutes a work made for hire, the author or authors of a work are the initial owners of the copyright that protects that work (*6*). As a result, an author may have the right to sue another author who copies his or her work without permission for copyright infringement. In addition, the author may decide whether and how to license or sell the copyright. Because authorship is of central importance to copyright law, the Supreme Court has established its own definition: "[a]s a general rule, the author is the party who actually creates the work, that is, the person who translates an idea into a fixed, tangible expression entitled to copyright protection (*7*)." In many ways, this legal definition of authorship is much broader than its ethical counterpart. In order to establish initial ownership of a work of authorship, there is no requirement that the author must approve the final version or attest to the accuracy and integrity of the work; for copyright purposes, all that matters is that the author helped write the paper.

There are many common pitfalls in determining authorship. Unofficial recognition of an author can take many forms but broadly involves recognizing as an author an individual whose contribution does not rise to the level of authorship. A prominent example is a courtesy or so-called "gift authorship," whereby a person is recognized as an author because he or she made possible, through funding or otherwise, the conduct of research or the preparation of a manuscript. Sometimes gift authorship may be awarded in an effort to increase the likelihood of publication or the credibility of the work. Awarding gift authorship is improper and widely discouraged—Yale University has even gone so far as to expressly forbid recognition of gift authors (*8*).

Gift authorship can present very real risks for the gift author since he or she is presumed to have reviewed and signed off on the work in question. The risks involved were made readily apparent in a recent situation involving an ecologist from Uppsala University in Sweden who wrote a high-profile *Science* paper on the effects of floating plastic on fish populations (*9*). After whistleblowers accused the author of fraud, there was an investigation, with the University ultimately finding that the published results were fabricated. The investigating body also concluded that the lone secondary co-author and supervisor also bore responsibility for the fabricated results. The co-author indicated he was "very disappointed" and had "very much trusted" the primary author. As a co-author, however, the fact that the he was "not very involved" in the study did not absolve him of responsibility. Because he had been listed as a co-author, he had to be held accountable for the fraudulent study, clearly demonstrating the potential reputational hazards of gift authorship (*10*).

### Defining Inventorship

While proper attribution of authorship has important implications for both ethical concerns and initial copyright ownership, inventorship is a similar concept with important ramifications in the law of patents, which provides legal protection for new and useful inventions (*11*). In some ways, the protection conferred by a patent mirrors that of a copyright. Just as the author is the original owner of a copyright, so too the inventor is the original owner of a patent, although an inventor may have a contractual obligation to assign his or her invention to another entity, e.g., an employer (*12*). One primary difference between the two is the subject matter that they can cover. In addition, unlike copyrights, patents provide their owners with the right to exclude others from making, using, selling, offering for sale or importing the claimed invention. Importantly, improperly designating inventorship in a patent can result in the patent ultimately being declared invalid or unenforceable (*13*). For these reasons, determining inventorship is a critical part of US patent law.

A patent includes a detailed description of the invention, typically referred to as the patent specification, which concludes with numbered sentences called "claims." In order to be an inventor, a person must show that he or she conceived of the "claimed invention"—referring exclusively to these numbered sentences (*14*). So it is entirely possible that an individual may contribute to subject matter in the detailed description of the invention section of a patent and may be considered an *author* of the patent specification, but because his or her contribution to the specification did not relate to the "claimed invention," that author might not be considered an *inventor*. Indeed, the drafting patent attorney may be viewed as an author of the patent document but rarely would be deemed an inventor. In any case, patent authorship effectively provides no legal benefits since patent ownership initially vests in the patent's inventors. Because of the complexities and potential legal consequences of improperly designating inventorship, consultation with a registered US patent attorney is highly recommended in determining who should be designated an inventor on any patent filing.

## Common Pitfalls

As the above discussion suggests, the everyday concepts of authorship and inventorship do not always line up with the official, ethical, or legal concepts. Moreover, the notions of inventorship and authorship can present special problems when it comes to NDAs. Below are several commonly encountered problems involving authorship and inventorship and NDA obligations.

### Competing Inventive Interests

Collaboration between individuals is ubiquitous in research and development. For individuals who are paid to invent, ownership of their inventions can be straightforward. Employment contracts typically require that employees must assign ownership rights to inventions discovered during their employment to their employer. In the absence of a formal employment contract creating such obligations, failure to ensure that all collaborators within a company have signed an appropriate NDA can be risky. In addition, a collaboration between individuals from different companies can prove problematic unless appropriate safeguards are implemented. The following hypotheticals demonstrate issues that can arise in these situations.

### *Hypothetical 1: The Rogue Intern*

Raincoats Corp. is developing a new waterproof material for use in raincoats. To protect their intellectual property, the company prepares an NDA for its employees. The company, having an interest in ensuring that the scientists' research remains secret, asks the scientists to agree to confidentiality by signing the NDA. But what if these scientists are not the only legal inventors? What if, for example, the company hired a summer intern to work on matters unrelated to the research project and neglected to have her sign the NDA? In her work at the company, the intern learned about the research project and had some ideas on how to improve on the waterproof material. Setting aside potential ethical concerns, in the absence of an NDA, the intern may be free to discuss the on-going research and the inventions themselves with third parties, including competitors, as desired, although such action may lead to potential claims against the intern rooted in tort law. In addition, depending on the degree of her contribution, she may even be entitled to file a patent application on aspects of the research resulting from her own inventions. In an extreme circumstance, she might even go so far as suing the research company for infringing *her* patent (*15*). Note, however, that if the intern's invention was created using company resources, the corporation may have a defense to any such infringement claim under the so-called "shop right" doctrine (*16*).

The above problem could have been avoided if the research company had done a better job of identifying potential inventors at the outset or by ensuring that all individuals associated with the project had signed an NDA. If the intern had been asked to sign the NDA as a condition of her internship, she likely would have been unable to legally seek patent protection for her invention. Also, the NDA could have ensured that all inventions derived from the project are owned by the research company. In this way, an NDA can provide two extremely valuable benefits for the company: (1) it protects against any undesired disclosure of the company's confidential information, e.g., inventions, and (2) it vests ownership in any developed inventions in the company.

### *Hypothetical 2: You Stole My Invention!*

Similar issues can arise when two companies informally work together or enter a joint development agreement (JDA) without sufficient contractual safeguards. Imagine a company, TireCo, invents a new type of tire having an improved tread pattern. TireCo realizes that the tire design works well, but could work better if the tire was manufactured using a new rubber material specifically adapted for the new tire design. They retain the services of Rubber Inc., a rubber company, which subsequently develops a new material according to TireCo's specifications. Without an NDA in place, Rubber Inc. may be entitled to file a patent application, without TireCo's permission, covering TireCo's new tire design manufactured with Rubber Inc.'s new material. But if the companies executed an NDA at the outset of the relationship, Rubber Inc. would not be permitted to file a patent application on the tire design even if it was made with its own material since doing so would necessitate disclosing TireCo's confidential information in violation of the NDA. Depending on the language in the NDA, Rubber Inc. still may be legally permitted to file a patent application on its new material, without limitations on its use, so long as doing so did not result in disclosing TireCo's confidential information. Ultimately, it comes down to the language in the NDA.

The above examples illustrate the importance of carefully outlining the parties' confidentiality obligations at the outset of the relationship. To this end, the agreement should carefully define the nature of the confidential information. In addition, the NDA should indicate which party would own any intellectual property that is independently or jointly developed under the contemplated relationship.

## The Timing of Publications versus Patent Filings

As discussed, authorship determinations are based largely on which individuals made a significant contribution to the work in question and agreed to share responsibility for its contents. These questions are not typically governed by NDA obligations, which, as a result, typically play little role in determining authorship. Instead, the primary consideration for NDAs in the context of publishing a work of authorship is in dictating *when* a work of authorship may be submitted or published. Thus, it is of critical significance that authors understand their NDA obligations when creating any new work of authorship.

Under US law, a patent may be obtained for any useful invention that is new and non-obvious (*17*). What is considered "new" and "non-obvious" is determined relative to the "prior art." Although the definition of "prior art" varies among countries, it generally includes all patents, printed publications and other publicly available information that existed prior to the filing date of the patent application (*18*). Although US law provides, with certain limitations, a grace period for filing a patent application within one year of an earlier publication, the laws of many other countries do not include such grace periods (*19*). Hence, as a general rule, a patent application should be filed before any works of authorship, e.g., articles or presentations, are submitted for publication or presented. This timing can be critical; the publication of an article before filing a patent application can result in refusal of the application or in any resulting patent being declared invalid.

Another consideration is ensuring that you have received proper permission to publish or patent any work related to the subject matter of the NDA. As discussed in greater detail below, NDAs vary in the information and activities they cover, and it is important to make sure that any publishing or patenting activity complies with the terms and scope of the NDA. However, even beyond the express scope of an NDA, it is important to recognize that publishing and patenting by their very nature can run afoul of an agreement to maintain confidentiality. That is, the acts of submitting (i) a patent application to the US Patent & Trademark Office, or (ii) a work of authorship for publication, could be deemed the public disclosure of confidential corporate information. As a result, any employee subject to an NDA obligation of confidentiality concerning his or her research should ensure that the owner of the confidential information, typically the employer, has agreed in advance to the filing of the patent application and/or submission of the publication so as to avoid an inadvertent breach of his or her NDA obligations. Such issues can typically be avoided by coordinating any planned publications, presentations or patent filings with the employer's intellectual property counsel.

**Understanding Your NDA**

Another common issue that can arise with NDAs is failing to fully understand the scope of the agreement, which can have broad impacts on inventorship and authorship in addition to creating areas of potential liability for the parties. While it is easy to conceptualize NDAs as simple agreements to keep secrets, there is actually a great deal of variety in the features that can be built into each NDA. Some common issues that can arise are outlined below.

*Unilateral and Bilateral NDAs*

One important feature relates to which parties to the NDA are obligated to maintain confidentiality. Generally, an NDA can be unilateral (sometimes referred to as a one-way NDA) or bilateral (sometimes referred to as a mutual NDA or a two-way NDA). The characterization of an NDA as unilateral or bilateral has implications on which party is contemplating providing the confidential information and which party has obligations to maintain the information as confidential.

A unilateral NDA contemplates confidential information going from one party to the other party, but not the reverse, and imposes an obligation of confidentiality only on the party receiving the confidential information.

*Hypothetical 3: Scaled-Up Production*

Assume Rubber Inc. from Hypothetical 2 completed development of its new rubber material for TireCo's tire design. Rubber Inc. then engaged ManuCo in connection with scaling up production of the new material. TireCo decides to enter into a unilateral NDA with ManuCo as part of its manufacturing contract. Under the terms of the NDA, ManuCo is obligated to keep the new rubber material, as well as the method of producing it, confidential. Rubber Inc. remains free to do what its information as it pleases, including potentially sharing that information with other manufacturers in attempting to identify another less expensive manufacturer.

Further, assume ManuCo develops confidential information concerning its scale-up process. If ManuCo chooses to share that confidential information with Rubber Inc., it does so at its own peril, since Rubber Inc. would not have obligations of confidentiality concerning ManuCo's information under the unilateral NDA. The unilateral NDA effectively places no obligations on Rubber Inc. but does create confidential obligations on ManuCo.

Under a bilateral NDA, on the other hand, the obligation would be on both parties. In our hypothetical, Rubber Inc. would be required to maintain ManuCo's confidential information as secret, just as ManuCo would be obligated to maintain Rubber Inc.'s information as confidential. The choice between a unilateral and a bilateral NDA often turns on the specific factual circumstances. Broadly, the form of the NDA reflects the nature of the information exchange. In many circumstances, parties insist on bilateral NDAs in the hopes of creating a more fair and balanced relationship. Regardless of which form is employed, once an NDA has been executed, it is important to fully understand which parties have obligations of confidentiality.

### *The Scope of Confidentiality*

NDAs are commonly prepared to facilitate discussions between individuals or other entities. In particular, in the JDA context, well-written NDAs are typically quite narrow in scope and are usually limited to cover only that confidential information that is necessary to be disclosed between the parties. As a result, NDAs should carefully define the scope of the confidential information that is contemplated under the agreement. A common mistake made by individuals operating under such NDAs is believing that since an NDA is in place, they can speak freely and disclose any confidential information without any repercussions. On the contrary, the disclosure of confidential information not contemplated by the NDA is likely not subject to the terms of the NDA, and the party receiving that confidential information is not obligated under the NDA to maintain that information as confidential. This can create significant issues, as explained in the following situation.

### *Hypothetical 4: Loose Lips*

Imagine TireCo and Rubber Inc. entered into a two-way NDA, and the contemplated scope of the NDA related to "materials for use in new tire designs." In the course of working together, the potential market for the tires was discussed by the parties, and an employee of TireCo shared a confidential list of possible customers with Rubber Inc. understanding that it was safe to do so under the NDA. Rubber Inc. then shared that confidential customer list with a competing tire company. Did Rubber Inc. violate the NDA? Likely not, since the customer list was not within the defined scope of the agreement.

*Hypothetical 5: Idle Chatter or Co-Invention?*

Now imagine that Bob, an employee of Rubber Inc., suggests a modification to the tire design to Jill, an employee of TireCo. Jill is impressed with the idea, tests it out, makes some minor modifications to the design and files a patent application naming herself as sole inventor covering the new tire design. Did Jill violate the NDA by filing the patent application? Likely not, since Bob's tire design modification was not covered by the scope of the NDA, which was limited to materials for new tire designs. Should Jill have named Bob as a co-inventor? Possibly. It really depends on the scope of claims in Jill's patent application and the degree to which Jill and Bob collaborated on the new invention. Does Rubber Inc. have a legitimate claim as a co-owner of the patent application? Again, it depends on the parties' agreements. But if the NDA was limited to "materials for use in new tire designs," any joint invention ownership provisions in the NDA may be similarly limited in scope—providing little support for Rubber Inc.'s ownership claim. Of course, if Bob should have been named as a co-inventor, he may have an ownership claim to the patent application as a co-inventor, which may support Rubber Inc.'s ownership claim if developing new tire designs were within Bob's scope of employment for Rubber Inc. In any event, this hypothetical illustrates some of the complex legal issues concerning inventorship that can arise when individuals from different companies share confidential information outside of the scope of the parties' NDA. The takeaway is this: before sharing confidential information, it is important to fully understand the scope of the NDA between the parties.

*Verbal Disclosures of Confidential Information*

Oftentimes individuals operating with the best of intentions mistakenly believe that, when possible, they should only disclose confidential information verbally so as to limit accidental disclosure to third parties. This approach can create significant issues depending on the language of the NDA.

Most NDAs include a verbal communications provision which requires documenting such verbal communications so as to ensure that the scope of what the parties consider to be confidential is clearly laid out. Although the language can vary widely, these provisions usually provide that any verbal communications of confidential information must be confirmed *in writing* within a certain period of time in order for the verbal communication to be considered as confidential information subject to the NDA. Failure to adequately document such verbal communications may give the impression that the verbally disclosed information is *not* confidential, resulting in its ultimate disclosure to third parties. As a result, it is important that individuals operating under an NDA understand the terms of the verbal communications provision of that NDA and follow up in writing as appropriate in compliance with the NDA.

## Conclusion

Determining inventorship and authorship can be difficult even under ordinary circumstances. Experienced authors and inventors and even attorneys often struggle in assessing whether a colleague's contribution rises to the level of authorship or inventorship. These issues can be compounded when obligations of confidentiality arise, potentially leading to complex legal issues such as breach of contract claims as well as ownership disputes. Individuals contemplating the exchange of confidential information under an NDA would do well to ensure that they understand the scope and terms of the NDA and consider potential problems that may be created by publishing or patenting subject matter that could be claimed to be another's confidential information. In view of the many pitfalls that can arise when operating under NDAs, questions concerning what can or cannot be shared under an NDA, as well as whether and when to publish or file for patent protection, should be addressed to appropriate legal counsel for clarification as needed.

## References


1. Yale University. *Guidance on Authorship in Scholarly or Scientific Publications*. https://provost.yale.edu/policies/academic-integrity/guidance-authorship-scholarly-or-scientific-publications (accessed May 10, 2018).
2. International Committee of Medical Journal Editors. *Defining the Role of Authors and Contributors*. http://www.icmje.org/recommendations/browse/roles-and-responsibilities/defining-the-role-of-authors-and-contributors.html (accessed May 10, 2018).
3. American Chemical Society. *Ethical Guidelines to Publication of Chemical Research*. http://pubs.acs.org/userimages/ContentEditor/1218054468605/ethics.pdf (accessed May 10, 2018).
4. International Committee of Medical Journal Editors. *Defining the Role of Authors and Contributors*. http://www.icmje.org/recommendations/browse/roles-and-responsibilities/defining-the-role-of-authors-and-contributors.html (accessed May 10, 2018).
5. *Subject matter of copyright: In general*. U.S. Code, Section 102(a), Title 17, 2012.
6. *Ownership of copyright: Initial Ownership*. U.S. Code, Section 201(a), Title 17, 2012.
7. *Community for Creative Non-Violence v. Reid*, 490 U.S. 730, 737, 1989.
8. Yale University. *Guidance on Authorship in Scholarly or Scientific Publications*. https://provost.yale.edu/policies/academic-integrity/guidance-authorship-scholarly-or-scientific-publications (accessed May 10, 2018).
9. Lönnstedt, O. M.; Eklöv, P. Environmentally Relevant Concentrations of Microplastic Particles Influence Larval Fish Ecology. *Science* **2016**, *352*, 1213–1216.
10. Enserink, M. Swedish Plastics Study Fabricated, Panel Finds. *Science*. **2017**, *358*, 1367 (This specific misconduct is further summarized in this source).
11. *Inventions Patentable*. U.S. Code, Section 101, Part II, Title 35, 2012.

12. *Beech Aircraft Corp. v. EDO Corp.*, 990 F.2d 1237, 1248 (Fed. Cir. 1993).
13. *PerSeptive Biosystems, Inc. v. Pharmacia Biotech, Inc.*, 225 F.3d 1315, 1319–21 (Fed. Cir. 2000); *Pannu v. Iolab Corp.*, 155 F.3d 1344, 1350 (Fed. Cir. 1998).
14. *REG Synthetic Fuels, LLC v. Neste Oil Oyj*, 841 F.3d 954, 962 (Fed. Cir. 12016).
15. *Ethicon, Inc. v. U.S. Surgical Corp.*, 135 F.3d 1456, 1462–64 (Fed. Cir. 1998).
16. *Wommack v. Durham Pecan Co.*, 715 F.2d 962, 965–66 (5th Cir. 1983).
17. *Conditions for Patentability*. U.S. Code. Sections 102-103, Title 35, 2012.
18. *Conditions for Patentability; Novelty*. U.S. Code. Section 102, Title 35, 2012.
19. *Conditions for Patentability; Novelty: Exceptions*. U.S Code. Section 102, Title 35, 2012.

## Chapter 6

# Case Study: The Difference Between Authorship and Inventorship

**Jeffrey Kovac***

**Department of Chemistry, University of Tennessee, Knoxville, Tennessee 37996, United States**
***E-mail: jkovac@utk.edu.**

This ethics case explores the difference between authorship and inventorship, but the principle ethical issues concern the relationship between a research advisor and a graduate student. Although the specific issue has to do with patent rights, this case can lead to a more general discussion regarding ethical issues that can arise in the advisor-student relationship.

## Patent Application

Bonnie Sanchez was a graduate student working in the research group of Arthur Taylor, a well-known synthetic polymer chemist. Taylor had an idea for a new catalyst for stereo-specific polymers which he asked Bonnie to develop in the laboratory. After a year of research with the usual frustrations, Bonnie got the catalyst to work and began exploring is usefulness in the polymerization of a variety of monomers. After seeing Bonnie's early results, Taylor was convinced that his catalytic system could be scaled up and would be useful in the polymer industry, so, he contacted the university patent office and filed a patent application listing himself as the sole inventor. He did not tell Bonnie about the patent application.

After Bonnie had shown that the catalyst was broadly applicable, Taylor asked Bonnie to write an article describing the research she had done. Because this was Bonnie's first article, there were multiple revisions, but eventually they submitted the article to *Macromolecules*, with Bonnie as the lead author. The published article received a great deal of attention from both academic and industrial polymer chemists, and both Bonnie and Dr. Taylor received calls and e-mails from companies interested in using the catalyst. As a result, Bonnie went to Dr. Taylor and asked, "Shouldn't we file a patent application?"

"I already have," replied Dr. Taylor. "I went to the university patent office before we wrote the article to make sure I had established inventorship."

"Wait a minute," said Bonnie. "Shouldn't I be listed as a co-inventor? After all, I'm the one who got the system to work."

Dr. Taylor patiently explained, "That is not how the system works. The inventor is the person who had the idea. Just because you were a co-author, even the lead author, does not make you a co-inventor. The rules are different for patents than for articles."

"That doesn't seem fair," said Bonnie, "I think I deserve some of the credit."

1. Was Dr. Taylor's behavior in not telling Bonnie about the patent application professionally responsible? Why or why not?
2. How could Dr. Taylor have handled the situation differently to avoid an unpleasant discussion with Bonnie?
3. Do you think that Bonnie should be listed as a co-inventor? Why or why not?
4. Assuming that the catalyst is commercially successful, should Bonnie receive any financial compensation?
5. What responsibilities do the chemistry department and the university's research office have in educating faculty, postdocs, and graduate students in the basics of patent law and the difference between inventorship and authorship?

## Chapter 7

# Misconceptions about Copyright and Permissions

**C. Arleen Courtney[1] and Eric S. Slater[*,2]**

**[1]Copyright Associate, American Chemical Society, 1155 Sixteenth Street, NW, Washington, DC 20036 United States**
**[2]Senior Manager, Copyright, Permissions, and Licensing, American Chemical Society, 1155 Sixteenth Street, NW, Washington, DC 20036 United States**
**[*]Email: e_slater@acs.org.**
**E-mail: a_courtney@acs.og.**

Over the years, the ACS Copyright Office has dealt with numerous misconceptions pertaining to Intellectual Property (copyright, trademarks, and patents). The purpose of this chapter is to dispel those misconceptions.

## Introduction

In keeping with the overall theme of this Symposium Series volume, it is of critical importance that authors in the scientific community have some familiarity with copyright and permissions. As stated in The ACS Style Guide, 3rd Ed., "Copyright law is a cornerstone of intellectual and scientific exchange." (*1*). While it is not required that one become expert in this area, it is imperative that authors obtain some degree of understanding, as most US publishers' copyright policies are based on United States Copyright Law.

For the purposes of this chapter, the focus will be on American Chemical Society policies, which, as stated above, are based on United States Copyright Law. It is possible that ACS policies may be identical or similar to that of other scientific publishers; however, ACS cannot speak for other publishers, and it is likely that policies can and do differ. This is due to their interpretation of the wording in the law.

Intellectual property consists of the following areas: copyright, trademark, patents, and trade secrets. For purposes of this chapter, the focus will be on copyright. That said, it is critical to respect all intellectual property, regardless of type, so as not to infringe on another party's work.

## What Is Intellectual Property?

Intellectual Property (IP) refers to creations of the mind, such as inventions; literary and artistic works; designs; and symbols, names, and images used in commerce.

IP is protected by law, for example, patents, copyright, and trademarks, which enable people to earn recognition or financial benefit from what they invent or create. By striking the right balance between the interests of innovators and the wider public interest, the IP system aims to foster an environment in which creativity and innovation can flourish (*2*).

Note that laws pertaining to these types of IP exist throughout the world. For purposes of this chapter, we are concerned only with United States law, and in particular, copyright law. It might be helpful to readers, though, to provide a brief primer on trademark law and patent law.

A trademark is a word, phrase, symbol, and/or design that identifies and distinguishes the source of the goods of one party from those of others. A service mark is a word, phrase, symbol, and/or design that identifies and distinguishes the source of a service rather than goods. Some examples include: brand names, slogans, and logos (*3*). The term "trademark" is often used in a general sense to refer to both trademarks and service marks.

Permission is required to use trademarked content, e.g., there are instances when authors want to use university or corporate symbols (logos) in their publications. Consistent with copyrighted content, permission from the trademark owner is required to reuse this material. Permission is also required to use the ACS and ACS journal logos.

A patent for an invention is the grant of a property right to the inventor, issued by the United States Patent and Trademark Office. Generally, the term of a new patent is 20 years from the date on which the application for the patent was filed in the United States or, in special cases, from the date an earlier related application was filed, subject to the payment of maintenance fees. U.S. patent grants are effective only within the United States, U.S. territories, and U.S. possessions. Under certain circumstances, patent term extensions or adjustments may be available (*4*).

## Copyright Basics

Copyright law is a doctrine of federal law and found in Title 17 of the United States Code. Currently, copyright law in the United States is governed by The Copyright Act of 1976. Note that all references made to copyright law in this chapter pertain to The Copyright Act of 1976 (hereinafter referred to the "statute").

Copyright is a form of protection provided by the laws of the United States to the authors of "original works of authorship" that are fixed in a tangible medium of expression. An original work of authorship is a work independently created by a human author and possesses at least some minimal degree of creativity. A work is "fixed" when it is captured (either by or under the authority of an author) in a sufficiently permanent medium such that the work can be perceived, reproduced, or communicated for more than a short time. Copyright protection in the United States exists automatically from the moment the original work of authorship is fixed (*5*). Therefore, unpublished works are copyrighted. They do not need to be published to be covered under copyright.

With regard to the above paragraph, there may some questions as to what constitutes a "minimal degree of creativity." While there is no written definition in the statute, the courts have weighed this issue. Essentially the accepted meaning is that works consist of content original to the author, i.e., content that the author creates on his or her own. (See generally, Feist v. Rural) (*6*). If authors are including content verbatim, adapted, or used in part that is not original to them taken from other sources, permission to use that content is required in most cases. Exceptions would include works that are in the Public Domain. This is consistent with ACS Policy, as it is based on the law.

Material created by U.S. Government employees is in the Public Domain and can be used without permission; however, credit should be given to the U.S. Government agency. Material published prior to 1923 is generally in the Public Domain; however, there are exceptions. For example, if a photo of artwork painted in the 1600s appears on a Web site, that photo is not necessarily in the Public Domain because it could have been taken recently.

Just because material appears on the Internet does not mean that it is available for use without permission. Someone owns copyright to that material, and permission is necessary unless it is otherwise stated on the Web site or there is a Creative Commons license attached to the material. Always check the site's Terms & Conditions to determine how the material can be used. Including Creative Commons material in an ACS journal article or book chapter is considered to be a commercial use of the material (not an educational or non-profit use); therefore, if the Creative Commons licensed or the Terms & Conditions do not allow commercial use, then the material cannot be used.

## What Can Be Copyrighted

Under Section 102 of the statute (*7*), copyrightable works include the following categories:

- literary works;
- musical works (including any accompanying lyrics);
- dramatic works (including any accompanying music);
- pantomimes and choreographic works;
- pictorial, graphic, and sculptural works;
- motion pictures and other audiovisual works;
- sound recordings;
- architectural works

Note that the category of literary works includes scientific works (e.g., journal articles and book chapters) and compilations.

## What Cannot Be Copyrighted

Many types of material are not eligible for copyright protection (*8*). Some examples include:

- Works not fixed in a tangible form of expression. Copyright does not protect ideas, only the fixed expression of ideas. Thus, a thought, not written down in any way, is not protected.
- Titles, names, short phrases, slogans, familiar symbols, or designs. (These items may be protected under trademark or service mark laws.).
- Lists of ingredients, contents, or facts.
- Ideas, procedures, methods, systems, processes, concepts, principles, discoveries, and devices. (These items may be protected under patent law.)
- Standard calendars, rulers, lists, or tables taken from the public domain and other works containing no original authorship.

## Who Is The Copyright Owner?

The copyright owner is the author or creator of the original work. An original work can be in the form of an article, photograph, illustration, figure, table, etc. Copyright does not protect ideas, only the actual expression of the ideas.

There are two exceptions when the author or creator of the original work is not the copyright owner: (1) when the copyright was transferred in writing to another person or entity (usually via a copyright status form or journal publishing agreement); or (2) when the work was created as a work-made-for-hire. In a work-made-for-hire situation, employees create the work within the scope of their employment; therefore, the copyright owner is the employer. The employer may be an individual, corporation, or university. The above two examples represent typical scenarios when publishing scientific or scholarly works. Works-made-for-hire can also cover, among other specified categories, independent contractor situations under certain conditions and if the parties have agreed in writing (*8*).

For ACS purposes, works-made-for-hire generally means that a researcher has authored an article for publication in an ACS journal as part of his or her employment: for example, at a university, or in industry. Under U.S. Copyright Law, the researcher's employer owns the copyright for that work under the work-made-for-hire doctrine. The employer is then required to transfer copyright to ACS under ACS Policy.

Examples of copyright owners are (1) the person who puts pen to paper or (2) the person whose fingers are on the computer keyboard and the work is saved on a hard drive, or (3) the person who's finger is on the camera shutter. (It is not the person in the photo unless the person in the photo took the picture; this includes "selfies".) That person is the copyright owner until he or she transfers copyright in writing to someone else, such as to a publisher. That work must be in a tangible medium such as a hard drive, on paper or canvas, or on film.

There are several methods of obtaining permission to use material for which the author is not the copyright owner.

1. Via the RightsLink permission system: Many publishers have their articles attached to the RightsLink permission system so that authors can locate at the publisher's Web site the article in which the requested material appears, and from there, go directly into the RightsLink permission system. Once an account is set up, permission can be obtained within minutes. There is a Quick Price Page so that requesters can see if a fee will be charged before the order is processed. If a fee is charged for the use of the material, there is usually a payment selection requesters can choose from in RightsLink. The Author would be advised to choose "Pay by invoice" instead of charging the fee to a credit card in case the author's publisher does not approve the use of the requested material. Authors should not assume that their publisher will reimburse the author for the fee.
2. Via forms provided by the author's publisher: The publisher's editorial office can provide forms for authors to use to request permission from other publishers or to send to copyright owners. It is recommended that the publisher's form be used instead of authors drafting their own permission requests because the wording on the publisher's form includes all of the rights that the author's publisher will need.
3. Via a form supplied by the publisher who owns copyright: Some publishers post an online permission request form for authors to use to request permission for material owned by that publisher.

## Duration of Copyright

Under current U.S. copyright law (*9*), length of copyright is life of the author plus 70 years. For works-made-for-hire, length of copyright is 95 years from the date of first publication or 120 years from the date of creation, whichever is shorter.

Works authored for ACS and most other scientific publishers fall into the work-made-for-hire category (with copyright being transferred to ACS and to most other scientific publishers), and are published, thus having a 95-year copyright term. The 120-year reference in the statue applies to unpublished works.

## Use of Photographs

For the use of photos not taken by the authors on a paper, the photographer signs a form granting permission and, if people appear in photographs, a model release form supplied by the author's editorial office must be signed by each person in the photo. If someone in the photo is deceased, the person's spouse or family member or the executor of the will can sign the model release. If a minor under the age of 18 appears in a photo, a parent or the child's guardian signs the model release.

If the photo is of the author on the paper, do not assume that your university or company owns copyright to that photo even if the photo was given to an author by the university or company to use. Check to be sure that the photographer either works for the university or company and taking the photo was done on a work-for-hire basis. If the photo was taken by a professional photographer, the photographer may still own copyright to the photo unless the photographer transferred copyright to the university or company who commissioned the photo.

## Stock Photos/Licenses

A license is needed to use photos from stock photo companies such as iStock, Shutterstock, Getty Images, etc. If an author wants to use the stock photo in a manuscript or on the cover of a publication, the wording in the license should cover the proposed type of use, the format in which the publication will be published (such as print and/or electronic format), and the license should be issued to the author. If required in the wording of the license, credit must be given to the stock photo company and to the photographer.

Some stock photo companies issue two types of licenses: one for educational or non-profit use and one for commercial use. Some photos are limited to educational or editorial use only. If you are submitting a manuscript to a publication that is to be sold (such as a purchased book or fee-based journal), that is considered to be a commercial use even if the publisher itself is not-for-profit (such as ACS).

If you already have a license for a photo issued to your university or company, unless otherwise worded, that license cannot be used in publications not published by your university or company. In other words, if you are submitting a manuscript to be published by a publisher that is not your university or company, a separate license for the photo needs to be issued by the stock photo company.

## Cover Art

When authors are asked to submit artwork for journal or book covers, if material not created by the author is used on the cover, permission is required and credit needs to be given on an inside page of the journal or book.

There is an exception to this general rule. For ACS journal and book covers, ACS editorial office staff has cover art forms to be completed and signed by the copyright owner, publisher, or photographer so that permission can be obtained. These forms are legal documents and need to be completed in their entirety and signed by the copyright owner, by the publisher, or by the photographer, not by the author. If the journal author created the cover art, the ACS Journal Publishing Agreement covers that artwork. If the book author created the cover art, a cover art form should be signed by the book author. The ACS Copyright Status Form does not cover artwork for book covers.

## Table of Contents/ and Abstract Graphics

When ACS authors submit a Table of Contents (TOC)/Abstract graphic for their manuscripts, there is no place to include a caption under the graphic so credit cannot be given to the copyright owner or publisher if the authors on the paper did not create the graphic. This would include Public Domain material such as a graphic created by a U.S. Government employee because credit cannot be given to the U.S. Government agency. Therefore, all portions of an ACS TOC/Abstract graphic must be created by the authors on the paper.

A TOC or Abstract graphic not created by the authors on the paper could, however, be used as a TOC or Abstract graphic if it is included as a figure in the manuscript as long as permission has been obtained from the copyright owner/ publisher and credit is included in the figure caption.

## Use of Material on the Internet

Never assume that anything that appears on the Internet can be used without permission.

Always look for the Terms & Conditions link on Web sites to determine under which conditions the material on that site can be used (such as print or electronic formats or for profit or non-profit use) and how to obtain permission. The Terms & Conditions link is often found at the bottom of the page where the words "Terms & Conditions" or "Copyright" appears but sometimes it is located elsewhere on the page. Click on that link and read the wording.

## When Is Permission Required?

Permission is required if the author does not own copyright to material whether it is used verbatim, adapted, modified, or used in part. Once permission is granted, credit must be given to the copyright owner. If the copyright owner does not supply the wording for the credit line, the author's publisher can supply wording for the credit line. ACS's standard credit line is worded as follows:

"Reprinted/Reproduced/Posted (or Adapted or Reprinted in part) with permission from [COMPLETE REFERENCE CITATION]. Copyright [YEAR OF PUBLICATION] American Chemical Society." Insert the appropriate information in place of the capitalized words.

See the ACS Style Guide at <pubs.acs.org> for sample credit lines.

## Copyrighted Material Used in the Classroom

Although Section 107 of the U.S. Copyright Law on "Limitation on Exclusive Rights: Fair Use" states that "the fair use of a copyrighted work, including such use by reproduction in copies or phonorecords (the term "phonorecords" refers to sound recordings) or by any other means specified by that section, for purposes such as criticism, comment, news reporting, teaching (including multiple copies for classroom use), scholarship, or research, is not an infringement of copyright" (*10*), it is recommended that permission be obtained to make photocopies of or to post on a password protected site articles or chapters for classroom use for students registered in a class.

This can be done via the RightsLink permission system, through the Copyright Clearance Center, or by requesting permission from the copyright owner or publisher. University Copy Centers or commercial copy shops (such as Kinko's) staff usually will not make photocopies of articles/chapters without seeing a copy of the permission.

## Use of Copyrighted Material in Theses/Dissertations

If a student is including non-text material that the student did not create or is not the copyright owner in a thesis or dissertation, permission is required and credit must be given.

Likewise, if the student is including non-text material that appeared in a submitted, accepted, or published manuscript of which the student was a co-author, the student should consult the publishing agreement that was signed for the manuscript for wording that covers that type of use.

If a student is including a portion of the text or the entire submitted, accepted, published paper for which the student was a co-author, the student should consult the publishing agreement that was signed for the manuscript for wording that covers that type of use. It may be necessary to obtain written confirmation from the editor who handled the paper to avoid potential conflicts with prior publication or embargo policies. Appropriate citation of the work should also be given.

## Fair Use

The Fair Use Doctrine (*10*) included in the U.S. Copyright Law is a very small section of that law compared with the wide use conception of authors who want to use the "Fair Use Doctrine" as a defense for not obtaining permission. **A good rule of thumb that we recommend to authors is that authors are never wrong when they ask for permission but they could be wrong if they do not obtain permission.**

## Guidelines for Choosing Non-text Material

From past experience we have found the items on this list of non-text material can be very difficult and or time-consuming to obtain permission and it is recommended that these items not be used:

- Currency (paper or coins) from any country.
- Maps (with the exception of maps created by Unites States Government employees).
- Postage stamps or flags from any country.
- Still photos or fictional characters from movie studios.

- Trademarked items such as company/university logos, images, and products (e.g., Coca-Cola, Honda, Rubik's Cube). These items are protected under trademark and permission is required.
- Use of Wikipedia, Flickr, and similar websites – material from these sites has not been vetted and, consequently, the information regarding the copyright status cannot be verified.
- Photos/slides of artwork or rare books from museums or libraries. Although libraries send photos from their collections for a fee, they are unable to grant permission to use photos if they are not the copyright owners. In such cases, the releases received as part of the invoices from libraries are not valid. Permission must be obtained from the photographer of the photo and the person in the photo must sign a Model Release unless the photo was taken prior to 1923.
- Material from the Internet should be used with caution. Someone owns copyright to that material, and it may not be clear from the information provided on the site who owns the copyright. For example, a photo of a painting by a Sixteenth Century artist appears on a Web site, such as Wikipedia, and the person who submitted the photo attached a Creative Commons license to it. That photo is probably not in the Public Domain, nor should a Creative Commons license be attached. That photo could have been taken yesterday by a photographer commissioned by a museum; therefore, the museum owns copyright to that photo. If that is the case, permission from the museum would be required.
- Photos, artwork, etc. from publications for which the author does not know the source, especially photos or artwork that appeared in "old" publications. Without knowing the source, it cannot be assumed that the photos or artwork are in the Public Domain merely because of their age.

## Non-text Material That Can Be Used without Permission

Material created or photos taken by U.S. Government employees is in the Public Domain. Credit should be given to the U.S. Government agency. Material created by contractors to the U.S. Government (such as Argonne National Laboratory, Pacific Northwest National Laboratory, and many other national laboratories) requires permission and credit. This material is not in the Public Domain.

It is not enough to include a reference citation for the copyrighted material used in a manuscript, nor is it enough to give credit and not obtain permission. If an author did not create the material or the author transferred copyright to a publisher, permission is required and credit must be given to the original source of the material, to the copyright owner, or to the photographer.

## Transferring Copyright to Publishers

Many publishers require authors to transfer copyright to their submitted manuscripts. This is done via a publisher's agreement such as the ACS Journal

Publishing Agreement (JPA), which is signed for journal manuscripts, or the ACS Copyright Status Form (CSF), which is signed for book chapters. Be sure to keep a copy of the form so that you are aware of the rights granted back to you as an author (*11*, *12*). ACS does not accept substitute forms or addenda.

The signatories and the form or section selected for the JPA or the CSF depends upon the affiliations of the authors on the paper, not on who supplied the funding for the research.

If **all** of the authors work for the U.S. Government, Form B of the JPA or the U.S. Government section of the CSF is signed by one of the authors. Those authors are not transferring copyright; they are confirming that all of the authors on the paper were bona fide U.S. Government employees.

If **all** of the authors work for the Government of Australia, Canada, or the UK, Form C of the JPA or the bottom of the CSF is signed by one of those authors. Those authors are not transferring copyright; they are confirming that all of the authors were bona fide employees of the Government of Australia, Canada, or the UK.

If neither of those forms are signed, then Form A of the JPA or the top portion of the CSF is signed by one of the authors who is not a U.S. Government employee or is not an employee of the Australian, Canadian, or UK Governments. It does not have to be the corresponding author who signs the form.

When the JPA or the CSF is signed, it means that the paper has been submitted on an exclusive basis and that the paper has not been submitted to another publisher. If the paper is rejected by the publisher, then the manuscript can be submitted to another publisher.

## If Authors' Names Are Added to or Removed from Manuscripts

If an author's name is added to or removed from a manuscript or an author's name has been changed (due to marriage or divorce), a letter addressed to the editor handling the paper needs to be signed by that author. The agreements signed by authors transferring copyright to a publisher are legal agreements and must reflect the legal names of the authors on a paper. Similarly, the name of an author on the publishing agreement must be the author's full legal name, not initials and the family name or the author's nom de plume.

## Deposit of Works to Institutional Repositories

ACS and many publishers permit authors to deposit their works into their institutions' repositories.

Such repositories are defined as online archives of universities, colleges, funding agencies, and other institutions and are key components of the emerging digital research infrastructure and can help the widest possible sharing of works. These repositories collect, preserve, and provide free, unrestricted online access to all types of institutional research outputs, seamlessly linking data, knowledge, and scholars (*13*).

## Full Disclosure by Authors to Publishers

Authors are required to disclose to publishers any prior publication issues and any other information that would have an influence on the publication of the manuscript.

Material that has been published prior to it being submitted to publishers needs to be divulged to publishers when the manuscript is submitted. Examples of prior publication are theses or dissertations, conference proceeding materials that have been widely distributed to attendees or have been for sale to non-attendees, reports that have been published, and material posted in certain repositories such as ResearchGate. ACS editors make the determination as to whether or not that material can be included in the manuscript.

## Computer Software Publishers Use To Detect Plagiarism

Many publishers use software to detect whether there is plagiarism pertaining to submitted manuscripts. ACS uses the "iThenticate" plagiarism detection software for this purpose (*14*).

Plagiarism is defined as follows: Somebody has presented the work of others (data, words or theories) as if they were his or hers and without proper acknowledgement (*15*).

It is also important for authors to be aware of self-plagiarism, and the fact that most publishers do not permit authors to submit works that have already been published. Detection of plagiarism and/or self-plagiarism can result in sanctions being taken against the author of a paper containing plagiarized material.

Authors should familiarize themselves with the so-called COPE Guidelines (Committee on Publication Ethics). COPE provides advice to editors and publishers on all aspects of publication ethics and, in particular, how to handle cases of research and publication misconduct. It also provides a forum for its members to discuss individual cases (*16*).

## Disputes Between Authors

If there is a dispute between authors on a paper, the ACS editorial offices do not become involved if the dispute is between university or company authors. It is up to the university or company to solve those disputes. Other types of disputes should be reported to the editorial office that handled the paper.

## Open Access Material

Open Access is the free, immediate, online availability of research articles combined with the rights to use these articles fully in the digital environment (*17*).

ACS and many publishers now have policies related to open access publishing, as well as journals dedicated to open access availability.

ACS features five components relative to its open access initiatives (*18*):

1. ACS AuthorChoice: Allows authors to publish open access articles via several types of licenses, including Creative Commons.
2. ACS Editors' Choice: Articles picked by ACS editors on a daily basis that are freely accessible.
3. ACS Author Rewards: Credits provided to authors that cover article publishing charges in ACS Omega.
4. ACS Central Science: A multidisciplinary journal with no fees to authors or readers.
5. ACS Omega: An open access, multidisciplinary, peer-reviewed journal with fast publication and low author publishing charges.

If you want to use material from an ACS open access journal article, you need to obtain permission. Before you request permission, you should check to see if the article is designated as "ACS AuthorChoice" or "ACS Editor's Choice".

At this time, you cannot use the RightsLink permission system to obtain permission for those open access articles (eventually the use of RightsLink will be possible). Instead, you need to contact support@services.acs.org with your request and include the following information:

- A link to the ACS article from which you wish to reuse content
- The portion of content you wish to reuse (e.g., number of figures, entire article for thesis)
- A description of where the content will be reused (e.g., name of journal, book title, thesis).

Please allow 7 business days for research and response to your request.

## Creative Commons Licenses

The use of Creative Commons (CC) license has become prevalent in scientific publishing, and is directly related to the concept of open access. ACS makes use of several CC licenses as part of its open access policies.

The following is the description of CC licensing from the CC website: Creative Commons helps you legally share your knowledge and creativity to build a more equitable, acceptable, and innovative world. We unlock the full potential of the internet to drive a new era of development, growth and productivity (*19*).

There are six different CC licenses to choose from, depending upon the type of use to be permitted:

1. Attribution – CC BY
   This license lets others distribute, remix, tweak, and build upon your work, even commercially, as long as they credit you for the original creation. This is the most accommodating of licenses offered. Recommended for maximum dissemination and use of licensed materials.

2. Attribution-ShareAlike – CC BY-SA
   This license lets others remix, tweak, and build upon your work even for commercial purposes, as long as they credit you and license their new creations under the identical terms. This license is often compared to "copyleft" free and open source software licenses. All new works based on yours will carry the same license, so any derivatives will also allow commercial use.
3. Attribution-NoDerivs – CC BY-ND
   This license allows for redistribution, commercial and non-commercial, as long as it is passed along unchanged and in whole, with credit to you.
4. Attribution-NonCommercial – CC BY-NC
   This license lets others remix, tweak, and build upon your work non-commercially, and although their new works must also acknowledge you and be non-commercial, they don't have to license their derivative works on the same terms.
5. Attribution-NonCommercial-ShareAlike – CC BY-NC-SA
   This license lets others remix, tweak, and build upon your work non-commercially, as long as they credit you and license their new creations under the identical terms.
6. Attribution-NonCommercial-NoDerivs – CC BY-NC-ND
   This license is the most restrictive of our six main licenses, only allowing others to download your works and share them with others as long as they credit you, but they can't change them in any way or use them commercially (*20*).

## Conclusion

Based on the information provided in this chapter and in keeping with the overall theme of giving credit where credit is due, it may be helpful for authors to put themselves in the shoes of other authors and creators. It stands to reason that all authors would want to be credited when their material is used. Copyright and permission go hand-in-hand with this notion, which, is in fact, governed by policy and the law.

## References

1. *The ACS Style Guide: Effective Communication of Scientific Information*, 3rd ed.; Coghill, A. M., Garson, L. R., Eds.; American Chemical Society: Washington, DC, 2006; pp 77.
2. *World Intellectual Property Organization: What is Intellectual Property?* http://www.wipo.int/about-ip/en/ (accessed February 16, 2018).
3. *United States Patent and Trademark Office: Trademark, Patent, or Copyright?* https://www.uspto.gov/trademarks-getting-started/trademark-basics/trademark-patent-or-copyright (accessed February 16, 2018).

4. *United States Patent and Trademark Office: General information concerning patents.* https://www.uspto.gov/patents-getting-started/general-information-concerning-patents (accessed February 16, 2018).
5. *US Copyright Office Circular 1.* https://www.copyright.gov/circs/circ01.pdf (accessed February 16, 2018).
6. *Feist Publications, Inc. V. Rural Telephone Service Co., 499 U.S. 340* (1991). https://www.law.cornell.edu/copyright/cases/499_US_340.htm (Accessed May 8, 2018).
7. *United States Copyright Law, Section 102. Subject matter of copyright: in general.* https://www.copyright.gov/title17/92chap1.html#102 (accessed February 16, 2018).
8. *The ACS Style Guide: Effective Communication of Scientific Information,* 3rd ed.; Coghill, A. M., Garson, L. R., Eds.; American Chemical Society: Washington, DC, 2006; pp 78.
9. *United States Copyright Law, Section 302. Duration of Copyright.* https://www.copyright.gov/title17/92chap3.html#302 (accessed February 16, 2018).
10. *United States Copyright Law, Section 107. Limitations on exclusive rights: Fair Use.* https://www.copyright.gov/title17/92chap1.html#107 accessed February 16, 2018).
11. *ACS Forms and Instructions for the Journal Publishing Agreement.* https://pubs.acs.org/page/copyright/journals/index.html (accessed April 16, 2018).
12. *ACS Overview of eBook Publishing.* https://pubs.acs.org/page/books/submission/overview.html (accessed April 16, 2018).
13. *SPARC Repository Resources.* http://sparc.arl.org/resources/repository (accessed April 16, 2018).
14. *iThenticate: Plagiarism Detection Software.* http://www.ithenticate.com/ (accessed April 16, 2018).
15. *Committee on Publication Ethics: plagiarism.* https://publicationethics.org/category/keywords/plagiarism (accessed April 16, 2018).
16. *Committee on Publication Ethics: COPE.* https://publicationethics.org/ (accessed April 16, 2018).
17. *SPARC Open Access.* https://sparcopen.org/open-access/ (accessed April 16, 2018).
18. *ACS Open Access.* http://acsopenaccess.org/ (accessed April 16, 2018).
19. *Creative Commons: What we do.* https://creativecommons.org/about/ (accessed April 16, 2018).
20. *Creative Commons: About the Licenses.* https://creativecommons.org/licenses/ (accessed April 16, 2018).

## Chapter 8

# Case Study: Authorship Issues and Conflict in the U.S. Academic Chemical Community

**Jeffrey Kovac**

**Department of Chemistry, University of Tennessee, Knoxville, Tennessee 37996, United States**
***E-mail: jkovac@utk.edu.**

This case raises the issue of plagiarism. A chemist has published an article in which he has copied sections of previously published articles without quotation or attribution. This case can be used to clarify what plagiarism is, how it can be detected, and the consequences of committing this unethical action.

## Familiar Words

Dr. Isabel Simpson, an assistant professor at Small State University, was scanning the table of contents of a leading journal of organic chemistry when she noticed an article by Professor Sam Darrow, an up-and-coming synthetic chemist, that was quite relevant to her current research. She immediately saw how the elegant synthetic methodology could be used in a synthesis that her current graduate student was attempting, but, as she continued to read, she was certain that she had read some of the sentences, and even whole paragraphs, before, although she did not remember where. She found it hard to believe that Darrow had committed plagiarism, but the words just seemed too familiar.

Over the next several days, the suspicion nagged at Isabel, and she began looking at related articles. After scanning about 20, she found the one she remembered. Reading it, she realized that Darrow had copied several whole paragraphs, as well as some individual sentences, in his paper. Although articles submitted for publication are routinely checked for plagiarism, this instance apparently had not been detected. She went back to Darrow's article to look for a reference to this paper but did not find one. She now had proof of her suspicion. What should she do?

1. Should Isabel contact Professor Darrow directly with her evidence of his plagiarism? If so, how should she do that: an e-mail, a phone call, or a letter?
2 Alternatively, should she contact the chair of his department or someone else at his university to discuss the situation?
3. Should she contact the editor of the journal in which the article was published?
4. Before doing anything, should she consult with an ethics officer at her university or a senior colleague to explore options?
5. Is this really a serious matter? Perhaps Isabel should just do nothing. After all, the science seemed right and a few plagiarized words did not make the paper any less useful. It is easier not to make trouble.

## Chapter 9

# Teaching Students Where Credit Is Due

## Two Lesson Plans for Teaching Documentation and Assignment of Credit

**Judith N. Currano***

**Chemistry Library, University of Pennsylvania, 231 S. 34th Street, Philadelphia, Pennsylvania 19104-6323, United States**
***E-mail: currano@pobox.upenn.edu.**

This chapter introduces two possible modules that can be used to teach the ethics of authorship to undergraduate and graduate students. The first module introduces the importance of documentation at a level appropriate for lower-level undergraduate students and consists of an interactive lecture and brainstorming session. The second module is a set of two classes introducing the ethical issues affecting authors. It is appropriate for upper-level undergraduates and graduate students, and it consists of a lecture and discussion and a set of case studies. Particular attention is given to those cases related to documentation and the assignment of credit to authors. Both modules have been successfully employed at the University of Pennsylvania.

## Introduction

Through the centuries, as technology has improved, access to information by the masses has also become easier. In ancient times, oral dissemination of information allowed a speaker to communicate only with individuals within reach of his or her voice, meaning that the information spread slowly and to relatively small groups of people at a time. Individuals who wanted to pass the news along would first need to learn it and then travel to the place where they wanted to share it. The written word improved the breadth of information dissemination since written works could be copied and shipped from one location to another. As literacy grew amongst the masses, the demand for written work increased, and mass-produced literature became possible with the invention of the printing press; this, combined with improvements in transportation, made it possible to share information much more broadly and quickly. Circulating libraries and, later, free libraries helped get the information into the hands of people of moderate income, and, for well over a century, most scholars have relied on mass-produced, printed books and journals to learn about the newest findings in their fields. As computers became more sophisticated, however, the tide of information dissemination began to shift again, and, by the dawn of the twenty-first century, most researchers were producing and relying on books and journal articles that had been written using word processing software and disseminated via the Internet. At present, many new journals are produced in an online-only format, and scholars enjoy the instantaneous access that the digital medium provides.

Scholars are not the only individuals who have enjoyed the ease of producing and consuming information in a digital environment. Just about everybody can easily prepare and share any information that they want, and a person walking down the street with a tablet or smartphone can quickly access anything from a scholarly journal article to the latest sports results. Media like Facebook and Twitter encourage people to pass along information that was produced by others, and an entire vocabulary has grown around such dissemination, such as "sharing", "retweeting", etc. Unfortunately, the rapid spread and accessibility of information have also engendered carelessness in evaluating what one has read, and erroneous or blatantly false information sits alongside quality facts. Digital information is also ephemeral; Web sites disappear, individuals decide to delete content that they had posted online, and, when information is updated, the previous version simply ceases to exist. Because of this, it is essential that people reference the source of the external information that they use; yet, social norms often ignore attribution completely. It is no wonder that students and young scholars are confused about what they need to attribute and how they should go about doing so effectively!

This chapter lays out two modules that can be used to teach students about the ethical responsibilities of scientific authors for attributing the information that they use. The first is for use with undergraduate classes; the lesson described was developed for an organic chemistry laboratory class but could easily be adapted for any course. The second introduces graduate students to the publication process and lays out the ethical responsibilities of authors in this process. Both have been successfully used for many years in the Chemistry Department at the University of Pennsylvania.

## The "Why", "What", and "How" of Documentation: An Undergraduate Lesson Plan

In the mid-2000s, the chemistry librarian at the University of Pennsylvania noticed that undergraduates taking the organic chemistry laboratory class, CHEM 245, had common questions involving the location of the physical properties of the substances used in their laboratory work. Working with the instructor, she developed a 50-minute lecture introducing methods of searching for published values of the physical properties of the substances that they used in the laboratory.

This lecture was delivered early in the semester, and, gradually, as the students became more self-sufficient, the questions subsided. However, scheduling permitted this lecture to be given only to those students who took the class during the spring semester; fall and summer students were out of luck. The instructor wanted all of the students to receive this training, so, she changed the timing of the lecture. Instead of holding it during the lecture portion of the class, she scheduled it for the first laboratory session, sitting alongside laboratory check-in. This lengthened the session from 50 to 90 minutes, and, since the class happened during check-in, every student was required to attend it. For the first several years, the focus of the session was on the structure of the chemical literature and methods of locating physical and chemical properties of substances. However, upon the request of the instructor, the lecture was expanded to include guidance on the importance of documenting the source of scientific information. The outline of the workshop currently given appears in Table 1.

**Table 1. Outline of Information Searching Workshop Taught to Organic Chemistry Laboratory Students at the University of Pennsylvania**

| *Topic* | *Details* |
|---|---|
| Outline of the Chemical Literature | • Brainstormed list of literature types<br>• Differences between primary, secondary, and tertiary literature |
| Introduction to the Handbook | • Definition of handbook<br>• Process by which a handbook is produced<br>• Quality indicators<br>• What happens when handbooks fail |
| Methods of Searching for Substances | • Chemical name<br>• Molecular formula<br>• Chemical structure<br>• Unique identifiers (ex. CAS Registry Numbers) |
| Example Search | • Mechanics of searching for a chemical in a sample database |
| Introduction to Documentation | • Why one cites others' work<br>• What needs to be cited<br>• How one cites others' work appropriately |

Although a discussion of the outline of the chemical literature may not seem like an ideal venue in which to introduce ethical issues, it can be a great time to begin discussing the importance of good documentation practices. Students brainstorm a list of information sources, and the instructor divides them into primary, secondary, and tertiary literature sources. Of all the genres that the students identify during the brainstorming session, they are probably most familiar with the tertiary, or review, literature; yet, it is the literature that is most diverse. Certainly, students tend to understand its purpose the least, far less than that of the primary or secondary literature. Stressing that the tertiary literature offers a scientist a starting point to research a topic by reviewing relevant literature for a particular purpose serves both to explain the limitations of this genre and to begin the discussion of documentation. A tertiary literature source without good documentation will not guide a researcher to further information on a topic and is, therefore, less desirable than one that is well-documented.

Explaining the purpose of a "handbook", a term with which students are increasingly unfamiliar offers similar opportunities to continue the documentation discussion. Handbooks of properties compile those values that the editors find to be the most useful and high-quality, and they present them to the reader in a concise and easy-to-use fashion. A well-designed handbook should also reference the primary literature in which the values were originally published. This helps a researcher to understand the conditions under which the value was observed and to determine whether or not experimental procedures varied between the primary reference and the researcher's own work. Despite the fact that the arcane term "handbook" is used to describe this type of literature, it is important to stress to the students that this genre can appear in printed or electronic format and can, in fact, more closely resemble a searchable database than a book of values.

The true meat of the ethical discussion, however, occurs in the section of the class devoted to documentation. Here, the focus is on the students' own work, the sources that they need to reference, and the styles that they should use when doing so.

### Why One Cites the Work of Others

The instructor can lead the students to a discovery of the importance of citation through the use of another brainstorming session. She begins the conversation with the question, "You know that it is important to cite others' work, but why is this the case?" Most students "raised" in the American educational system will answer that the practice of omitting appropriate documentation is plagiarism, and they know that "plagiarism is bad", but they have not devoted much thought to the critical need for documentation in scientific research. The reasons for documentation, or the "why" of documentation, can be broken down into three categories: the "altruistic" reason, the "practical" reason, and the "self-preservation" reason, and the instructor continues the discussion, asking appropriate leading questions until all three purposes have been voiced.

The altruistic reason for citation is to give credit to the originator of an idea, a statement, or a value. Most students understand that they are "giving credit where credit is due," but they are less familiar with how critical this is in academic chemistry. Talking about the "altruism" of documentation can preface a discussion of the use of impact metrics to determine the value of individuals and publications, if desired, and a mention of how these metrics are used in academia. Job retention, promotions, and awards can all be linked to the impact that a scientist has had on his or her field, and impact is currently measured in citations; hence, citations are a form of "currency" in academia.

It takes slightly more fishing to get students to define the practical reason for citing material, which is to allow readers to get more information about something that will help them understand the work at hand. In the case of physical and chemical property searching, it helps them to learn more about the experimental conditions and other, related properties that were observed at the same time and can hint at the interconnectedness of properties. Once the students arrive at this purpose of documentation, it quickly becomes clear to them that referencing allows an author to write a more concise paper, focusing on the parts of the research that are novel and referring the reader elsewhere for peripheral or past work.

Oddly enough, the "self-preservation" reason for citation is usually the one that the students mention last. When asked, many will respond, "If you don't cite others' work, it's plagiarism," or, "You're not giving credit to someone else, so, you're stealing their work." It takes a little more effort to convince them that, in the process of not giving credit and stealing words or ideas, an author who fails to cite research on which they rely is also taking full responsibility for the veracity (or lack thereof) of the undocumented work. By documenting the work that they use, authors are able to indicate the source of their own conclusions, and, if the origins of a paper's conclusions are understood, the paper may remain useful even if past work upon which it relied is later found to be inaccurate.

### What Needs To Be Cited?

Once students understand the reasons for citation, the discussion turns to the items that actually need to be cited. This part of the class begins with a paraphrase of the University of Pennsylvania's formal definition of plagiarism: "using the ideas, data, or language of another without specific or proper acknowledgment" (*1*). Students are encouraged to cite material they use whenever they are in doubt about whether or not it is necessary. On the whole, students are aware that they need to cite written concepts that they read in texts, but they are not as clear about what they should do with property information and reaction mechanisms. They are instructed always to cite property values and then asked why this is a good practice. Leading questions asked by the instructor can help them discover that property values may depend on experimental protocols or be calculated from other observed properties; therefore, a single property may have many reported values, even within the same source. By citing the source, a writer can direct the reader to additional information about the experiment or equation that determined the property at hand.

Reaction mechanisms are a little trickier, and this is a good opportunity to introduce the concept of "general knowledge". The instructor first asks the students what the term "general knowledge" means. Answers returned usually include, "something that everyone in the field knows" or "an unambiguous fact". While this may indeed be "general knowledge", for the purpose of citation, the students are asked to think of general knowledge as something that they, as experts in the discipline, were able to pull out of their own memories without the use of external sources. Reaction mechanisms provide a useful example. If a student writes a mechanism from memory, it is considered part of his or her "general knowledge". If he or she verifies that the known mechanism is correct using a source, citing that source is optional. If, on the other hand, the student does not know the mechanism by heart and looks it up before reporting it, then he or she is required to cite the source in which the answer was found. Likewise, if the student writes the mechanism, looks it up to verify it, and changes what he or she has written based on the information in the source, the student should cite the source. The discussion of "general knowledge" closes with a brief discussion of "optional citation". If a student knows a fact but wants to prove that others agree that this fact is true, he or she should locate and cite an authoritative source for the fact.

The last topic discussed in this section is when to cite the primary literature and when to cite databases of property values. The instructor performs a sample search for the melting point of aspirin using Reaxys, which reports more than fifty reported values for this property from a variety of journal articles and patents. She chooses one value and directs the students' attention to the reference to the primary literature beside the value in the database; then, she asks the students how many of them would cite Reaxys. A few will raise their hands. She then asks how many students would cite the primary article. More students tend to raise their hands, and she asks them to keep their hands up. Then, she asks any student who has read the primary article to keep his or her hand up, and all of the students lower their hands. She tells them that the students who still have their hands read are allowed to cite the primary article; everyone else must cite Reaxys. The key take-away is that students may only cite information that they themselves have read. This portion of the class closes with a brief discussion of the reasons that one might choose to cite the primary article over the database and vice versa.

## How Should Information Be Cited?

The last section of the class is extremely brief; students are given a handout containing examples of journal articles, books, free databases, and fee-based databases cited according to the conventions described in the *ACS Style Guide* (*2*). Since the most recent version of the *ACS Style Guide* was published in 2006 and does not have good guidelines for the citation of non-journal electronic information, the database style is a locally-generated amalgam of the ACS styles for citing several different information types. The instructor points out that, when citing electronic information, one should always include the date on which the information was accessed. Print books, continue to exist even after a new edition is published, and as long as one includes the publication date and edition number, the information in them can be found and verified for as long as the book is available. Electronic information sources, on the other hand, tend to replace content as new information becomes available. Therefore, the value of a property that appears in an update of an electronic source may be completely different than the same value was before the update. Also, electronic information can completely disappear if a work is sold or withdrawn by its author or publisher, and URLs can change even for works that continue over time. By including the access date, a student is "proving" that the information existed at that location in that moment of time.

## Ethics in Scholarly Communication: A Graduate Lesson Plan

The University of Pennsylvania offers a course in chemical information, CHEM 601, which is required of all Ph.D. students and elective for master's degree and undergraduate students. While most of the curriculum deals with methods of locating and evaluating information, the last two-class module, introduced in 2006, focuses on ethics in scholarly communication. The focus of the module is on the ethical responsibilities of individuals at each stage of the publication process, and documentation is one topic of conversation. The first session of the module follows a manuscript through each stage of the publication process and lists the job duties and the ethical responsibilities of each participant in the process. The second session presents a series of case studies dealing with scholarly communication and documentation. Prior to the first session, the students are asked to read *On Being a Scientist* (*3*) from the National Academies' Press and three chapters of the *ACS Style Guide* (*2*): Chapter 1, which deals with ethics in scholarly communication; Chapter 6, which covers peer review; and Chapter 7, which includes information about copyright. To ensure that the students have read the material before class, the instructor requires the students to complete an online pre-test, highlighting the important concepts from the reading.

### Session 1: Lecture and Brainstorming on the Ethical Responsibilities of Authors, Editors, Reviewers, and Readers

The first session begins with the instructor tracking a hypothetical manuscript through the publication process and explaining what happens to it at each stage. The framework and key talking points come primarily from Chapter 1 of the *ACS Style Guide* (*2*) but also includes information provided by the other required readings and the instructor's personal experience with the process. The manuscript must first be written by the authors, who adhere to a set of practical and ethical norms while writing. It then goes to the editor, who, following ethical norms associated with his or her role, must determine whether or not it is appropriate for the journal and assign it to reviewers. The reviewers, also behaving ethically, recommend publication and make suggestions for improvement, and the editor makes a final decision about the publishability of the manuscript with or without revisions. If the manuscript is accepted, it becomes the responsibility of the production team, who prepare it for publication, and the author approves the final copy. Finally, it lands in the hands of readers, who have the responsibility of reporting errors that the author or editor must address.

After this discussion, the students divide into groups to discuss the ethical responsibilities of the four main entities involved in the publication process: authors, editors, reviewers, and readers. The students are asked to make a list of all ethical issues that could face individuals in each role and, when their lists are complete, to write them on the section of the board labeled with the appropriate role. All groups are asked to comment on each role, placing a check mark beside anything on their list that another group has already written on the board. Figure 1 shows a diagram that could result from this part of the discussion.

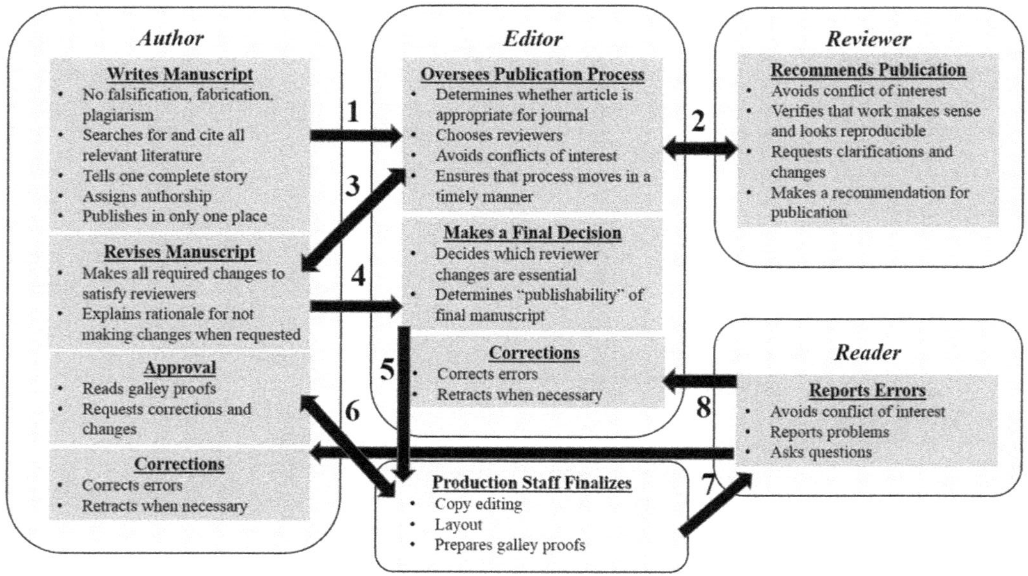


*Figure 1. Diagram indicating the path of a manuscript through the publication process and some of the ethical responsibilities of each party in the process, based in part on Chapter 1 of The ACS Style Guide (2).*

After the students finish posting the results of their conversations, the instructor brings the class back together to discuss the items on the board. The discussions of the editor's and the reviewer's jobs are fairly rudimentary and are more informative than discussion-based since the students will not assume these roles until later in their careers. However, the discussions of the author and the reader should be reasonably well developed since the students are already filling these roles. The discussion of the author's duties and responsibilities focuses on things to do and avoid to be an ethical author, and some key points appear in Table 2.

**Table 2. Topics That Can Be Included in a Classroom Discussion of Ethical Authorship Practices***

| *Things to Do* | *Things to Avoid* |
|---|---|
| Perform a comprehensive background search | Fabrication, or inventing experiments, publications or data |
| Determine appropriate authorship and author order | Falsification, or choosing and omitting data to support a conclusion |
| Prepare a clear, accurate, reproducible account of research | Plagiarism and failure to request permission to use previously published material |
| Report the origin of any data not original to the project | Dual publication or submitting to multiple journals simultaneously |
| Give credit to work that has come before | Conflict of interest or appearance thereof |
| Declare financial interests | Unnecessary fragmentation of research into many publications |
| Choose an appropriate journal for publication | |
| Work in a timely manner | |
| Correct errors and retract if necessary | |

* Information is inspired by and adapted from the appendix to Chapter 1 of the *ACS Style Guide* (*1*).

Most of the concepts in the list make sense intuitively to the students. However, it is important to spend some time discussing the concept of dual publication and "self-plagiarism". This is a good time to tell the students that, depending on the details of the publication agreement signed by the authors, publishing a paper in multiple journals may be illegal, in addition to being unethical. If the authors transferred copyright to the publisher, then publishing the same article in a second location violates copyright law. The intellectual property no longer belongs to the author, and the author is not permitted to use it for any purpose outside of the rights reassigned to the author in the copyright transfer. The Committee on Publication Ethics (COPE) has some very clear and well-written information on self-plagiarism available on their Web site; this can be used as supplementary reading for the students before or after the class discussion (*4*).

Another concept that interests the students is that of authorship, and most classes contain at least one individual who knows someone who was party to an authorship dispute of some sort or another. There are many gray areas about which individuals should have authorship of a paper and which individuals should simply be acknowledged. Most scientific entities agree that an author is an individual who has made a significant intellectual contribution to a project and who has participated in the writing of the paper, but there is a great deal of room for interpretation of what constitutes a "significant intellectual contribution". Facilities specialists, undergraduate research assistants, and laboratory technicians can be used as good examples of individuals who may perform the same service and qualify for authorship in some circumstances but not in others, depending on the nature of their contribution and the main thrust of the work to be published. This usually makes for a lively conversation with much difference of opinion.

**Session 2: Discussion of Case Studies**

The second session of the module is devoted to two sets of case studies, one focusing specifically on authorship and peer review, and the other devoted to these and other ethical areas in scholarly communication. Two of the case studies in the first set were written specifically for this purpose by the instructor, while the rest were assembled from a variety of published sources, including *On Being a Scientist* (*3, 5*) *Scientific Integrity* by Francis Macrina (*6*), and a pamphlet on responsible conduct of research published by the Association of American Medical Colleges (*7*). The students receive the case studies at the end of the previous session and are asked to meet with their groups between classes to discuss the various scenarios. Each group is assigned responsibility for leading the discussion on one or more of the cases. At the beginning of the second session, the groups meet for five or ten minutes to organize their thoughts before the discussion begins.

A list of case studies used in the class appears in Table 3, and those cases that include issues related to documentation and the assignment of credit are highlighted in bold in the table and discussed in greater detail in this chapter. In the students' packets, the cases appear in no particular order, although the first set contains cases related solely to authorship and peer review. The students are encouraged to read and digest each case carefully, with the hope that they will draw parallels between cases for themselves. During class, however, the instructor groups cases that deal with similar topics for ease of discussion. Cases 1.1, 2.1, and 2.2 all deal with attribution of information and are discussed as a set. Cases 1.2 and 1.3 deal with assignment of authorship and credit and are therefore discussed together. Cases 1.4 and 2.3 round out the discussion of authorship aspects by raising the issues of corrections and retractions (1.4) and industrial sponsorship and non-disclosure agreements (2.3). Case 2.4 focuses on the ownership and stewardship of data and can be discussed after Case 2.3 to build on concepts of data ownership introduced there. Case 1.5 is not related to the others; this case deals with peer review and can be discussed at any point, although if a primary investigator (PI) claims credit for a review written by his postdoctoral fellow, that can be considered plagiarism. The remainder of this chapter will focus on the first two case sets; the other cases, while being interesting for a variety of reasons, do not relate to documentation and assignment of credit and are therefore outside of the scope of this chapter.

### *Cases Involving Plagiarism*

The first set of cases discussed forms a unit related to plagiarism and appropriate documentation of the work and ideas of others. Case 1.1 focuses on the idea of plagiarism. It consists of three parts, and students are asked to identify whether or not each mini-case constitutes plagiarism and determine what an administrator or journal publisher should do if the article described was submitted for publication. In the first part, a researcher paraphrases material from a source. Because she did not directly quote it and she is only using it in the introduction of her paper, she does not include the reference in her bibliography. She has clearly plagiarized, and this part of the case serves as a good warm-up by allowing the students to reiterate the definition of plagiarism and indicate that the location of the plagiarized material within the article is irrelevant. In the second part of the case, a researcher uses a single review article to provide all background information for the introduction of his paper, documenting it appropriately. This is not plagiarism, but it is still sloppy research and possibly even misconduct because the researcher is relying on the review's author to interpret the primary source material, and he has not read the original articles himself. This hints at the fact that he has not done a complete literature search before beginning his research and is not giving proper credit to those who came before. In the third case, the researcher copied information from one of his previous papers verbatim into a new paper. This is an issue known as "self-plagiarism", or "text-recycling", and it offers the opportunity to continue the discussion of copyright transfer and dual publication that was begun in the first class.

The idea of self-plagiarism is continued in Case 2.1. This case comes from Macrina's *Scientific Integrity,* and the protagonist is the editor of a book who discovers that a prominent contributor has reproduced a figure from a journal article that he authored. The discovery occurs after the book has gone to press, so, no corrections can be made to the text of the book (*6*). If this case study is discussed directly after Case 1.1, the students are already primed to think about issues of text-recycling. Asking them what the author could do to be ethical in a case like this allows for a fuller discussion of the need to request permission to reuse figures and text from published works to which one does not own the copyright and the need to document the first publication of any reprinted figures or text to which one does own the copyright. Given that the figures appear in a journal article and a book, it is also possible to discuss the timeline for publication of each and the meanings of terms like "in press" and "submitted" when citing information that has been submitted for publication but has not yet been published.

**Table 3. List of Case Studies Used in the Second Class of the Graduate-Level Module on Ethics in Scholarly Communication***

| *Case #* | *Summary* | *Source* | *Ethical Issue(s)* |
|---|---|---|---|
| | | Case Studies, Set 1 | |
| **1.1** | **Identifying plagiarism** | **Original** | **• Plagiarism**<br>**• Sloppy background research**<br>**• Text recycling** |
| **1.2** | **Determining authorship and reporting the results of multiple experiments** | **Original** | **• Authorship and undergraduate laboratory assistants**<br>• Accurately reporting yields |
| **1.3** | **How many publications are enough?** | ***On Being a Scientist*, 1995, p. 15 (*5*)** | **• Fragmentation** |
| 1.4 | When does authorship responsibility end? | On Being a Scientist, 2009 (*3*) | • Correcting errors<br>• Uncooperative coauthors |
| 1.5 | Asking a post-doctoral researcher to review an article that has been assigned to one | Teaching the Responsible Conduct of Research Through a Case Study Approach (*7*) | • Confidentiality in peer review<br>• Plagiarism |
| | | Case Studies, Set 2 | |
| **2.1** | **Researcher publishing the same data in two different articles** | **Scientific Integrity, *3rd Ed., p. 87 (6)*** | **• Dual publication**<br>**• Timeline for publication** |
| **2.2** | **"Scooping" a graduate student** | ***On Being a Scientist,* 1995, p. 12 (*5*)** | **• Documenting unpublished work**<br>**• Protecting one's unpublished work** |
| 2.3 | Industrial sponsorship of academic research | On Being a Scientist, 1995, p.9 (*5*) | • Nondisclosure agreements<br>• External support |
| 2.4 | Who owns the data? | *Teaching the Responsible Conduct of Research Through a Case Study Approach* (*7*) | • Data ownership<br>• Data stewardship<br>• Importance of memoranda of agreement |

* Items highlighted in bold indicate issues surrounding documentation or assignment of credit.

Unpublished results form the bulk of the discussion of Case 2.2. This scenario, from *On Being a Scientist*, is a situation in which the students may well find themselves as they begin to present papers at national meetings and conferences; therefore, it bears some in-depth conversation. In the case, a student delivers a paper at a national meeting and subsequently holds a detailed conversation about his methodology with a prominent researcher in his field. The researcher proceeds to publish a paper that appears to depend on the studcnt's method but does not cite the communication (*5*). Discussion of the case begins with the correct method of citing unpublished work and personal communications and progresses to things that the student can do to protect his unpublished results while still communicating openly with other researchers. Some potential solutions to this problem include checking with the PI before the conference to determine the level of detail to use in discussions and keeping a "paper trail" of communications by holding detailed discussions via e-mail instead of verbally. Finally, the students are asked to consider situations in which the "scooping" scientist has acted ethically, including the parallel development of methodology. This allows one to explain how a well-documented and dated laboratory notebook can be used to resolve disputes over originality and intellectual property ownership.

*Cases Involving Assignment of Credit*

Assignment of credit in the form of authorship is covered in Case 1.2. In this scenario, a graduate student has had limited success in performing a step in her synthesis, so, she assigns this part of the project to her undergraduate laboratory assistant. He uses her methodology and achieves success only when he is forced to change suppliers for a key reagent because it is out of stock at the preferred supplier. The questions associated with this case prompt the students to think about whether or not the undergraduate should have authorship privileges on the paper that results from these experiments and the best way of reporting the range of yields obtained using the two batches of reagent. The students tend to be divided as to whether or not the undergraduate has made a "significant intellectual contribution" to the project, so, the instructor can lead them to some possible methods of making the undergraduate student's contribution more "author-worthy" by including him in higher-level decision-making and experimental design earlier in the process.

The discussion of authorship can be continued into Case 1.3. While the case itself deals with fragmentation, a potential solution to the problem can involve a discussion of author ordering. A young assistant professor and her students are conflicted about the number of publications that should result from a project. The professor thinks that a single paper in a prominent journal is sufficient, while the students are lobbying for two papers, likely to allow each of them to be a first author (*6*). Some talking points can surround the need to make publication decisions about the number of papers and the journals to which to submit earlier in the research process so that all participants agree before they reach the writing stage. This discussion can be broadened to touch on interinstitutional collaborations and the desirability of written memoranda of understanding, one of the topics under consideration in Case 2.4, which focuses on the ownership of research data (*7*). To "help" the professor solve her dilemma, the instructor could also suggest that the students share first-authorship, although joint first-authorship is a concept of which many graduate students may be ignorant. Introducing the idea of joint first-authorship opens other interesting avenues of discussion, including the maximum number of coauthors who can reasonably or ethically share first-authorship and is likely to engender lively discussion and questions on the mechanics of reporting joint first authors.

## Conclusion

The lesson plans presented in this chapter are intended to serve as examples of possible methods an instructor could use to teach students the importance of documentation in scholarly communication and ethical norms of assigning credit to contributors in scholarly work. The discussion of the "why", "what", and "how" of documentation is presented here in the context of an information skills lecture that teaches students to locate physical properties of chemicals, but it could just as easily be used to preface the assignment a term paper or other oral or written work that relies upon external sources. The "Ethics in Scholarly Communication" module was designed for use in a class on searching and using the chemical literature, but it can easily be slotted into a professional skills seminar class or, in fact, any upper-level chemistry class with a writing or peer-review component.

The case studies presented in the graduate lesson plan represent examples of cases that could be used to demonstrate the concepts discussed, but the sources cited in this chapter contain additional cases that could also be considered. Many other sources of case studies also exist. "The Lab" (*8*), produced by the Office of Research Integrity in the Department of Health and Human Services, is an interactive video dealing with a case in which a student's name is applied without her consent to a paper containing falsified data. Students can view the situation from the point of view of four different characters, and each decision point offers the potential for group discussion or individual reflection. Columbia University also has a fairly robust set of modules dealing with authorship and peer review (*9*), as does the National Academy of Engineering (*10*). Here, Caroline Whitbeck has assembled a collection of short cases on responsible authorship that can be used to start a conversation. In addition to these, she has also compiled a set of four cases dealing specifically with plagiarism (*11*). Plagiarism is a major problem facing all scholarly publishers, so, it is not surprising that many of them, including ACS Publications, the Society for Industrial and Applied Mathematics (SIAM), and Elsevier, have developed their own guidelines and instructional modules for authors (*12–14*). Finally, the text *Responsible Conduct of Research*, by Shamoo and Resnik (*15*), contains chapters on authorship and publication and peer review, each of which has a number of useful case studies at the end.

The goal of the modules presented in this chapter, as well as the materials cited above, is to ensure that the scholars of tomorrow are well-versed in the ethical norms of scholarly communication and documentation. Students need to understand the reasons for citing past work early in their course of study, so that, as they embark upon their research careers, they appropriately acknowledge past advances and thus perform their ethical responsibilities as authors.

## References


1. *Handbook for Students: Ethics and Original Research.* https://provost.upenn.edu/uploads/media_items/ethics-handbook.original.pdf (accessed May 1, 2018).
2. *The ACS Style Guide: Effective Communication of Scientific Information*, 3rd ed.; Coghill, A. M., Garson, L., Eds.; American Chemical Society: Washington, DC, 2006.
3. *On Being a Scientist: A Guide to Responsible Conduct of Research*, 3rd ed.; National Academy Press: Washington, DC, 2009. http://www.nap.edu/catalog/12192.html (accessed May 1, 2018).
4. *Text Recycling Guidelines*, https://publicationethics.org/text-recycling-guidelines (accessed May 1, 2018).
5. *On Being a Scientist: Responsible Conduct in Research*; National Academy Press: Washington, DC, 1995; https://www.nap.edu/catalog/4917/on-being-a-scientist-responsible-conduct-in-research-second-edition (accessed May 1, 2018).
6. Macrina, F. L. *Scientific Integrity: Text and Cases in Responsible Conduct of Research*, 3rd ed.; ASM Press: Washington, DC, 2005.
7. Korenman, S. G.; Shipp, A. C. *Teaching the Responsible Conduct of Research through a Case Study Approach: A Handbook for Instructors*; Association of American Medical Colleges: Washington, DC, 1994.
8. *The Lab.* https://ori.hhs.gov/TheLab/TheLab.shtml (accessed May 2, 2018).
9. *RCR Responsible Authorship and Peer Review.* http://ccnmtl.columbia.edu/projects/rcr/rcr_authorship/case/index.html (accessed May 2, 2018).
10. Whitbeck, C. *Responsible Authorship.* http://www.onlineethics.org/cms/research/modindex/auth.aspx#content (accessed May 2, 2018).
11. Whitbeck, C. *Scenarios on Plagiarism.* http://www.onlineethics.org/Resources/TeachingTools/Modules/19237/resethpages/plagiarism.aspx (accessed May 2, 2018).
12. ACS Publications. *Ethical Guidelines to Publication of Chemical Research.* https://pubs.acs.org/userimages/ContentEditor/1218054468605/ethics.pdf (accessed May 2, 2018).
13. Society for Industrial and Applied Mathematics. *Authorial Integrity in Scientific Publication.* http://www.siam.org/books/plagiarism.php (accessed May 2, 2018).
14. Elsevier. *Policies and Ethics.* https://www.elsevier.com/authors/journal-authors/policies-and-ethics (accessed May 2, 2018).
15. Shamoo, A. E.; Resnik, D. B. *Responsible Conduct of Research*; Oxford University Press: Oxford, 2014.

## Chapter 10

# Case Study: Teaching Responsible Authorship Practices to Graduate Students

**John D'Angelo**

**Division of Chemistry, Alfred University, Alfred, New York 14802, United States**
***E-mail: dangelo@alfred.edu.**

Nearly all students understand that plagiarism is not allowed. What is less clear to many is the issue of self-plagiarism. By the end of one's graduate career, a firm understanding of this must be possessed. Herein, cases of self-plagiarism are explored.

## Imagine This...

You are reading a paper to prepare for a research project you are starting, and being a studious worker, you read many articles as part of your literature survey. Suddenly, you get to an article that makes you stop and say to yourself, "I read this one already!" Shuffling through your stack, you find the paper that you remembered, and you see that they are, in fact, **not** the same paper. The authors are the same, the titles are very similar, and, when you glance through the two documents, you find that they largely identical and were published in the same year.

*Some points to ponder:*

- Is this bad for science since the same authors, after all, are reporting the results both times?
- What actions, if any, should you take?

Similar situations have actually happened. Recently, the *Journal of the American Medical Association (JAMA)-Otalaryngology-Head and Neck Surgery* was compelled to investigate one of its papers, by Maria Schietroma and his colleagues. This paper was found to have many similarities to a previous article that had been retracted because some statistical results were incorrect, causing the data to fail to support the conclusion. In the past, Schietroma had two papers retracted due to similarities to other published work. He defended the article currently under fire, claiming that the errors in the earlier (retracted) paper were corrected. He also claimed that the similarities between the two papers are due to the genuine similarity between the studies.

*Some additional points to ponder:*

- Do prior instances of misconduct (borderline or overt) warrant extra scrutiny of the authors in future questionable cases?
- Since the manuscript in question is similar to a retracted paper, are concerns of self-plagiarism nullified?

## Source

Stern V. Retraction Watch. *JAMA journal calls for formal investigation into surgery group's work.* https://retractionwatch.com/2017/12/28/jama-journal-calls-formal-investigation-surgery-groups-work/ (accessed May 4, 2018).

# Authorship in Collaborative Research Projects

# Chapter 11

# A Roadmap to Successful Collaborations between Primarily Undergraduate Institutions and Research Institutions

**David Rovnyak[1] and George C. Shields*[,2]**

**[1]Department of Chemistry, Bucknell University, 701 Moore Avenue, Lewisburg, Pennsylvania 17837, United States**
**[2]Provost's Office and Department of Chemistry, Furman University, 3300 Poinsett Highway, Greenville, South Carolina 29613, United States**
***E-mail: george.shields@furman.edu.**
**E-mail: drovnyak@bucknell.edu.**

Collaborating with colleagues at research intensive (R1) institutions is a fast growing trend in research at primarily undergraduate institutions (PUIs). As such partnerships are being initiated at a remarkable pace, it is timely to discuss emerging norms and best practices that can help these two very different research cultures build successful and lasting relationships. This article describes a chronological roadmap for all stages of a PUI/R1 collaboration. Emphasis is placed on strong communication between partners, with a focus on elucidating mutual expectations and leveraging the respective strengths of the PUI and R1 partners.

## Introduction: The Journey

The continuing growth in research at primarily undergraduate institutions (PUIs) is leading to a relatively recent area of fruitful opportunities for researchers at PUIs: working with colleagues in research-intensive institutions such as Ph.D. granting institutions, medical research facilities, government laboratories, and private institutes.

There are certainly a variety of precedents for PUI researchers working beyond the walls of their institutions, such as through sabbaticals and leaves on which a PUI professor may travel to another institution. However, this article addresses a growing trend in which the research environment at the PUI itself is involved collaboratively with R1 institutions. We have written previously about the important role of PUI research in the advancement of knowledge (*1*, *2*), and PUI/R1 collaborations represent one of the important mechanisms by which PUI research contributes to the scientific enterprise.

This article aims to provide insights and guiding principles that can apply to a wide variety of PUI/R1 collaborations. Even if a specific situation is not treated here, some universal themes emerge that should be applied to every situation. The main thrust of this guide is to urge the strongest models of communication among all participants. Openly discussing research goals, funding needs, publication expectations, and more will define the success of these partnerships. This article also aims to review some of the strategies that are used across all types of PUI settings.

Perhaps the greatest reason for emphasizing transparent interactions is that PUI/R1 collaborations are just beginning to gain a foothold in United States-based research, and therefore norms for PUI/R1 interactions are being developed from the ground up. Whereas research-intensive institutions have already evolved frameworks to enable collaborations, covering areas such as data sharing, intellectual property, and more, the key concept presently underlying PUI/R1 interactions is that two very different research cultures are now starting to learn how to forge successful partnerships. This article, therefore, also seeks to promote best practices that will help define norms for future PUI/R1 partnerships. It should be appreciated that there is no 'one size fits all' model for PUI/R1 collaborations. We aim in some cases only to raise the questions that should be asked, ultimately leaving the answers up to the participants. (Figure 1)

*Figure 1. Effective collaborations between PUI and R1 groups reflect a complex synergy between research cultures; with effective communication and planning, the relative strengths of these institutions can be leveraged to advance research goals and outcomes.*

Some key trends in the current landscape of US-based research can be identified that are driving the formation of PUI/R1 partnership. For example, whereas R1 institutions face an increasing mandate to shift their focus to solving more applied research questions, PUIs have a lower-stakes research culture that has more flexibility to tackle basic research questions. The complementarity of these environments is one driving force for bringing PUI and R1 groups together. For instance, when GCS was involved in a research project with a medical school, he was able to adapt his basic research using computational chemistry focused on development of transition state analogs for ester hydrolysis to the design of haptens that would potentially hydrolyze cocaine (*3–6*).

Another interesting trend in US-based science has been a gradual increase in working across research silos to accelerate the achievement of research goals. This rise in interdisciplinary collaboration at research institutions has stimulated R1 investigators to consider the availability of collaborative opportunities with PUIs. Not only may a PUI laboratory have the expertise and resources to fill critical needs in a research team, but PUI researchers have a culture that is supportive of faculty who delay research productivity in order to develop other areas of expertise. For example, when DR was approached about initiating NMR metabolomics work with the Obesity Institute at Geisinger Research, he and his lab spent almost two years intensively training, tooling up, and performing a mock clinical study of Type II diabetes before engaging in *de novo* clinical work. This effort developed new expertise but, echoing the theme of this section, also resulted in basic research on metabolomic sample handling with some surprising results on metabolite stability (*7*). The broader outcomes of developing new expertise to support research are compelling. Several students participated in the metabolomics work; as of this writing, Broc Wenrich is concluding a medical degree and assigned a residency, Phillip Martin is an NSF Graduate Research Fellow at PSU pursuing epidemiology in agriculture, and Matthew Miele resides in the Proteomics Core at Memorial Sloan Kettering Cancer Center.

As most researchers know, funding for US-based science has changed. Lower funding rates and gaps in funding for principal investigators at research institutions means that researchers receive less support for graduate students, post-docs, staff scientists, and technicians but still face high expectations for research productivity. It is increasingly recognized that research groups at PUIs consist of many undergraduate students who acquire nontrivial expertise and skillsets in the course of their work. Indeed, we and our colleagues often describe some of our undergraduate researchers as performing at the level of a first- or second-year graduate student. Forging PUI/R1 collaborations can help to mitigate the shortage of resources to ensure progress on research questions, while providing additional valuable research opportunities to undergraduate researchers.

The rewards of PUI/R1 collaborations are tremendous. Simply put, such partnerships provide more resources and expertise to both the PUI and R1 labs to solve a greater range of research questions at an increased pace. A PUI/R1 collaboration can make it possible to tackle research areas that neither could do alone. The professional development of both PUI and R1 researchers is enhanced through an expanded scientific network and through greater exposure for the work from more conference presentations and publications. A PUI/R1 collaboration can be a strategy for starting student-driven research programs in PUIs where research is less common, due to the shared resources and external validations of the PUI's abilities that this collaboration brings. Building authentic PUI/R1 relationships can also improve bids for new equipment and infrastructure at both institutions. GCS used his collaborations with medical schools to justify his need for increased computer resources, which he obtained for the MERCURY consortium through an NSF-MRI grant (*8*).

A further example illustrates the efficacy of PUI/R1 collaborations in rapidly developing research ideas. DR's publication in 2011 of the first correct exposition of an under-appreciated opportunity for sensitivity enhancement in NMR, involving solely undergraduate coauthors (*9*). This paper generated an unexpected degree of discussion in conference settings and publications and also led to the finding of other groups who were making similar discoveries. DR and a network of R1 groups developed a series of papers that solidified and deepened the theoretical and experimental basis for this phenomenon (*10–12*). This work, involving several undergraduate coauthors, has been cited over 140 times as of this writing, and has also resulted in a noteworthy undergraduate honors thesis for Melissa R. Palmer, who earned first-authorship on three papers (*12–14*).

Not only do these collaborations create additional opportunities for undergraduate students to develop as researchers, but the graduate students and postdocs in the R1 lab also benefit in several ways, such as through gaining mentorship experience and having more exposure to the PUI setting and the many rewarding career paths at PUIs.

Primarily undergraduate institutions vary considerably in their culture for student-driven research, the programs available for research, and in the levels of support available for research. These are serious issues, but we have seen very productive PUI/R1 collaborations arise across the diverse spectrum of PUI settings. Our experience is that any PUI/R1 interaction is greater than the sum of its parts and we urge colleagues to pursue such opportunities regardless of the particular environment for research at their PUI.

This article has been arranged with a chronological view, with anecdotes and examples given along the way. It can be read in order, or one can skip to a section that discusses a particular stage of most relevance to a given situation.

## Is Your Car Ready for the Road?

Time and money are the major impediments for all researchers, at every institution. Institutional support, or lack thereof, is a major problem for many regional state institutions and private liberal arts colleges without large endowments. Let's face it: there are a lot of different cars on the road, from BMW's to minivans. But any institutional environment has the potential for research excellence, and research excellence is necessary if a PUI faculty member is going to have a productive collaboration with an R1 faculty member. The Council on Undergraduate Research (CUR, see CUR.org) is the leading professional organization dedicated to helping faculty establish and maintain active undergraduate research programs. CUR has published a document, Characteristics of Excellence in Undergraduate Research (COUER), which CUR uses as a basis for their Campus-wide Award for Undergraduate Research Accomplishments (AURA). CUR has given the AURA award to nine institutions over the past three years, and the six PUI institutions who have won this award are Allegheny College (2015), The College of New Jersey (2015), Furman University (2016), the University of Wisconsin-Eau Claire (2016), Hope College (2017), and Denison University (2017). Of these six, Allegheny, The College of New Jersey, the University of Wisconsin-Eau Claire, and Hope College are all constrained by time and money, yet all have reached the top levels of undergraduate research excellence as defined by COUER and CUR. We guarantee that PUIs will continue to win this award!

Silverberg, Tierney, and Cannon have written about establishing vigorous undergraduate-driven research programs at institutions that are resource-limited, including discussing the importance of collaboration to achieving research goals and effecting cultural shifts (*15*). We pursued this further with Silverberg, who credited the crucial role of building PUI/R1 relationships in developing his large undergraduate research group at the Schuylkill campus of the Pennsylvania State University. Moreover, Silverberg points out that one does not necessarily have to engage in extensive collaborations or to connect with a major funded grant of a particular R1 lab; instead, he sought smaller, more focused interactions with R1 colleagues with short time frames and well defined outcomes that allowed him to achieve research goals early. Silverberg stresses the importance of approaching potential collaboration from the perspective of building relationships with R1 groups first, where not all need to become collaborations; many relationships that did not become collaborations still led to important speaking and networking opportunities and donations of equipment, including some from industry partners that were (and still are!) crucial to growing his research. In other words, in resource-limited environments, developing PUI/R1 collaborations is an effective strategy to getting more resources and changing the research culture of that institution.

In 2008, Armstrong Atlantic State University did not have a full-time summer undergraduate research program, yet, even though the science faculty taught a 4-4 teaching schedule, the faculty in the College of Science and Technology (CST) were active researchers who wanted to excel in undergraduate research. An active, 8-10 week summer program, where the students work full-time in the lab, is essential for faculty to maintain research productivity and for students to progress. Removing some of the time and money barriers allowed for the development of the first ever, summer undergraduate research program, despite the onset of the Great Recession (*16*). The Armstrong faculty also started an early "introduction to research" program, modelled after Hamilton College's program (*17*) but tweaked for the less-prepared students and more time-challenged and underpaid faculty, and it dramatically increased retention and graduation rates in science and engineering programs (*18*). This model also increased research productivity, as students worked multiple summers and academic years in the same research group, thus enhancing the ability of the Armstrong faculty to do real science. The Chemistry Department was a major driver of the success, with Will Lynch as the Department Head and Delana Gajdosik-Nivens as an Assistant Dean. Delana eventually became Dean of CST. The Armstrong success was so great that, when Armstrong was consolidated into Georgia Southern in 2018,

Delana became dean of Georgia Southern's College of Science and Mathematics, and Will became Department Head of Georgia Southern's chemistry department.

## Filling Up the Tank: Initiating PUI/R1 Collaborations

Research partnerships between PUI and R1 laboratories arise in many ways. For example, it is common for a faculty member at a PUI to maintain ties with colleagues from their former Ph.D. and postdoctoral laboratories that can grow into collaborations. An important point is that the frontiers of science are so advanced and require such extensive training that it is much more common and even expected to stay closer to one's Ph.D. and postdoc training when embarking on an independent career. The expertise that a PUI faculty member can contribute to joint projects with their prior laboratories or with colleagues who have moved to their own universities is one particularly natural mechanism for collaborations to arise.

An example of these close connections is the history of Jim Phillips, Professor of Chemistry at the University of Wisconsin-Eau Claire. UWEC is a relatively resource-constrained institution that won the 2016 AURA award for its excellence in undergraduate research. Jim, a graduate of Middlebury, obtained his Ph.D. in Ken Leopold's spectroscopy lab at the University of Minnesota, and was a NOAA postdoctoral fellow with Veronica Vaida at the University of Colorado. As a young faculty member at UWEC, he found himself needing to learn some computational chemistry to interpret the results of his experiments, and he has benefited greatly from a collaboration with Chris Cramer at the University of Minnesota. Cramer's willingness to be helpful and supportive had a multiplicative effect on Phillips' productivity, and were career changing. Indeed, GCS has experienced the same support from Cramer, who has given three presentations and provided valuable advice to PUI students and faculty at the MERCURY conferences. Phillips is an active and valued member of MERCURY, one of the few members who has both experimental and computational research efforts underway in his lab.

Presenting work and attending conferences remain among the best pathways to initiate collaborations, yet faculty at PUIs face significant challenges to attend conferences as they generally have limited access to funds to support travel, while teaching duties severely restrict the ability to attend conferences scheduled during the academic year. For example, for many years there has been a shifting trend among faculty at research intensive institutions towards attending smaller, more disciplinarily-specific conferences to help maximize interactions and visibility with other researchers who have shared expertise. Shifting to more focused meetings has also been driven in part by the increasing depth of expertise in research fields. As research activities increase at PUIs, we are observing that a similar trend is now taking root amongst PUI researchers who attend, with their students, a variety of smaller, more focused meetings for the same. While major national meetings still have important roles, we advise PUI faculty to ensure that they also target some more intimate, disciplinary-specific venues that promote strong interactions. For a long time, GCS took his computational chemistry group to the Sanibel Symposium in St. Augustine, FL, and later St. Simons Island,

GA, as the small size of this meeting meant that his undergraduates had the most networking chances and could improve their science through interactions with the research faculty in attendance. Similarly, DR and his students regularly attend ENC's (Experimental Nuclear Magnetic Resonance Conference) for valuable NMR immersion and are more frequently attending discipline-specific meetings (Chirality, SMASH, PANIC, EAS, etc.).

Conferences serve another major role. As a general norm in chemistry and related fields, most conferences do not separate undergraduate posters into a separate category. It is certainly very important for the development of the undergraduate student presenters that they defend posters in the same sessions as all other laboratories, but when PUI and R1 labs are presenting in the same sessions they also gain opportunities to discover shared interests and complementary expertise. One notable example was when Katrina Lexa presented her senior thesis research at the Sanibel Symposium in March of 2005. Katrina presented her work in the general poster session, where she had maximum attention from the R1 faculty, and received the most outstanding undergraduate poster award. Carlos Simmerling at Stony Brook and Adrian Roitberg at Florida offered advice on a new molecular dynamics technique, which allowed us to solve the problems we were facing and resulted in the design of a potential new anti-cancer drug (*19–22*). Katrina went on to earn her Ph.D. at Michigan and is now a research scientist at Denali Therapeutics. Katie Alser picked up the project after Katrina graduated, and she herself earned a Ph.D. in organic chemistry at Duke, and is now a research scientist at Eli Lilly. This project itself was a collaboration with a group of Albany Medical College researchers led by Tom Andersen (*21*).

The **M**olecular **E**ducation and **R**esearch **C**onsortium in **U**ndergraduate computational chemist**RY** (MERCURY, see mercuryconsortium.org) has been holding annual conferences at the end of the summer undergraduate research session for faculty and their undergraduates to meet R1 faculty, learn about career opportunities, and network with like-minded researchers. The conference has been held every summer since 2002 and averages about 100 attendees. MERCURY currently consists of 27 PUI faculty at 25 institutions, and many of these institutions have much less institutional support for research because they are regional public institutions like UWEC, Cal Poly San Luis Obispo, Central State, James Madison University, Monmouth University, Truman State University, University of North Georgia, or are private institutions with small endowments, such as Adelphi University, Iona College, Roanoke College, Siena College, and St. Edward's University. The conference serves as a source of support for all of the MERCURY faculty, as well as faculty who attend the meeting from other PUIs. Graduate students who attend learn about the PUI world, and undergraduates are able to view their own research through a larger research prism.

A final word on conferences is that we empathize strongly with the strain it puts on PUI faculty to stretch precious travel funds and organize student presentations and posters to attend conferences, but the visibility that this builds for the laboratory's research is a solid investment to stimulating future collaborations.

Collaborations often arise very organically when PUI students and faculty visit research institutions to use specialized equipment. For example, DR has led groups of students to Pennsylvania State University, Univ. of Pennsylvania, and Univ. of Delaware to take data, and his students have attended workshops at the Univ. of Connecticut Health Center as well. These visits have built relationships that led to fruitful collaborations. Additionally, grant-seeking exposes one's research expertise to large review panels of accomplished scientists around the country and further builds awareness of the potential collaborations a lab could offer. Jason Cody, at Lake Forest College, returns regularly to his Ph.D. institution, Northwestern University, to use facilities there. Karen Almeida at Rhode Island College, regularly takes advantage of core facilities, such as biophysical instrumentation, at the University of Rhode Island and Brown University. Costs for using these facilities is often lower for PUI institutions, as federal agencies try to encourage research with undergraduates.

Importantly, PUI faculty should be aware that their expertise and collaborations are needed because R1 labs are becoming overtaxed and collaborations between R1 labs are not as easily realized. A broader collaborative base is critically needed in the United States, and PUIs are playing a bigger part in meeting that need.

An encouraging example comes from the lab of Brent Feske, who, while an Assistant Professor at Armstrong, wrote an NSF-RUI grant on biocatalysis with biology professor Scott Mateer, which was funded in 2009. The two presented their RUI research at a Gordon Conference on biocatalysis. While at the conference, Feske approached Andreas Bommarius, a professor in the School of Chemical and Biomolecular Engineering at Georgia Tech. Feske suggested to Bommarius that, since they were in the same state and had very similar research goals, perhaps they should collaborate. Feske's NSF grant writing and reviewing experience had given him the vocabulary needed to be convincing when he explained the two ways that he could bring value to Bommarius's research group and projects. The first idea, on Intellectual Merit grounds, was that, because Bommarius was an engineer, his students typically struggled to make enzyme substrates with any type of chemical synthesis. Feske had established a successful track record of finding and developing synthetic strategies for synthesizing putative enzyme substrates that undergraduates could perform. He also was able to support the Georgia Tech group with chemical instrumentation trouble shooting. The second idea, on Broader Impacts, was that Armstrong's chemistry department, and Feske's group in particular, had been very good about collecting and processing all sorts of broader impact-related statistics for Armstrong science students (Underrepresented Minorities, Women, First Generation, along with programs that prepared students for graduate school). The PUI and R1 faculty both agreed that a collaboration could only strengthen a proposal coming from the R1. Since that meeting they have submitted four grant proposals together, two of which are funded. Feske was added to serve on a graduate thesis committee for a research student in the Georgia Tech group. These two Armstrong faculty, with busy teaching schedules and few institutional resources, developed a strong research program were doing high quality research with undergraduates, and had their own NSF-RUI grant that established them and gave them real credibility. In the newly-formed Georgia Southern University that has integrated the Armstrong Faculty, Feske is an Associate Professor of Biochemistry and Associate Dean of the combined College of Science and Mathematics. Mateer is Associate Professor of Biology and Associate Chair of the combined Biology Department.

Oftentimes, it is published work by a PUI faculty member, read by a R1 faculty member, that stimulates a research collaboration. GCS has had two productive collaborations based on published work that filled a gap for the R1 faculty member. The first was on cocaine catalytic antibodies, arose from work on ester hydrolysis and publication of transition state analogs (TSAs) (*3*, *4*). This attracted the attention of Don Landry at Columbia University College of Physicians and Surgeons, who was interested in designing haptens that mimicked TSAs as potential cocaine addiction treatments. This partnership resulted in three papers (*5*, *6*, *23*), and undergraduates Ed Sherer and Gordon Turner were major drivers of this project. Both attended graduate school, at Minnesota and Yale, and now work at Merck and Novartis, respectively. A series of papers on the structures and stabilities of water clusters by undergraduates and a high school student interested in atmospheric chemistry, led by undergraduates Meghan Dunn, Mary Beth Day, Tom Morrell and Kaye Archer (*24–30*), resulted in the prediction of all the relevant hexamers of water at various temperatures in 2011 (*31*). This paper caught the eye of microwave spectroscopy expert Brooks Pate at the University of Virginia, whose research team had just completed the first experimental determinations of the water hexamer and found all three of our predicted hexamers present in his experimental beam. This led to a highly productive and ongoing collaboration that has resulted in five papers, including two in *Science* (*32–37*). Meghan attended graduate school at Colorado, and now works at the EPA. Mary Beth completed a Ph.D. in Earth Sciences at the University of Cambridge. Tom went to Princeton for graduate studies, while Kaye is currently at Pittsburgh pursuing her Ph.D. degree.

While we have written mostly about how to stimulate collaborations, it is critically important to conclude this section with emphasis on the importance of saying no. Both of the authors of this article have been in a position to consider potential collaborations that exceed the time and resources available and have had to decline with regrets multiple times. For example, DR had declined previous offers for metabolomics collaborations at times when his lab could not make the commitment. It should be clearly recognized from exploratory initial conversations whether the scope of the collaboration is appropriate for the PUI laboratory an is worth the time of the PUI faculty member. R1 labs need to move fast, and if the PUI lab is not in a position to keep up, then it is better to say no in advance rather than suffer through the awkwardness of a relationship in which they cannot deliver on their half of the research plan. In the water cluster collaboration with Pate, described above, if Shields had not had NSF-RUI grant money to support a postdoc or research scientist (Berhane Temelso), then his lab would not have been able to move quickly enough to keep up with the world-class Virginia experimental group.

Prior to formalizing a collaboration, the PUI lab could offer to conduct limited exploratory or feasibility studies to determine the suitability of the work for the PUI lab. We are familiar with situations in which the R1 lab also proposes to do more feasibility studies to help aid the collaboration. Such work should always be distinguished as low-stakes work that occurs independently and prior to establishing a collaboration. As we will discuss in more detail below, planning carefully for resources and commitments takes on heightened importance in PUI/R1 collaborations, partly because such collaborations often involve substantial direct effort from the PUI faculty member.

Standard advice follows for any collaboration: that it not be forced but rather represent mutually beneficial resources, expertise, and goals to improve the chances of succeeding in a research endeavor.

## Hitting the Road: Formalizing the Collaboration

### Overview

A broader lesson borne from experience is that PUI/R1 collaborations should have carefully defined parameters and goals. Participants should aim to build a foundation based on recognizing the research cultures at each institution. Avoid the temptation to assume that there will be mutual understanding on a variety of logistics, but, instead, discuss up-front a complete set of logistical items spanning from funding realities to communication strategies.

Although this is the stage at which the collaboration will likely proceed, this crucial planning phase may yet reveal previously unanticipated barriers to performing the work, and it is far better to discuss this early than to assume that 'things will work out'. For example, DR had reached advanced planning stages of a structural biology collaboration with a large R1 group. Reilly Sonstrom, an accomplished undergraduate student at Bucknell who had developed expertise in protein expression as co-first author on a protein project (*38*), had been tapped to start extensive preliminary experimentation on this collaboration, when it was discovered that the system presented significant expression challenges that were anticipated to require postdoctoral effort and ultra-high fields for NMR. The decision was made between the labs to abandon this collaboration. Sonstrom went on to coauthor another work that featured her protein expertise (*39*) and went on to pursue a Ph.D. with Brooks Pate at UVA.

Both labs in a PUI/R1 collaboration are encouraged to reach a mutual understanding that the collaboration is more than outreach. One particular pitfall to avoid is a situation in which the R1 partner views the collaboration solely as an outreach venture, such as part of a broader impacts effort, and not as part of an active research project.

Recognize that not all PUIs are the same and that research institutions are equally diverse. It can be very constructive for partners to talk openly about the norms of research at their institutions. Points of discussion can include learning how investigators are reviewed by their institutions, the models they use for supervising their research groups, the modes of communication on which they rely (lab meetings, email, video chats, other), and even whether or not each institution allows identification cards for collaborators to facilitate visits and sharing library and other resources.

**Talk about the Science**

It is axiomatic that the goal of a collaboration is to pursue a research question with combined resources and expertise that improve the chances of and shorten the timeline for success. Put simply, all parties should aim to collaborate on mutually agreed scientific goals and questions.

At the risk of over-generalizing, we observe that successful PUI/R1 collaborations operate around clearly-defined objectives and criteria. These collaborations arise to meet very specific needs to advance scientific inquiry. Of course, research is inherently non-linear, and it is common to 'shift gears' to pursue unexpected findings, as well as to adapt to findings, but it is still essential to outline clear mileposts and specific aims.

It should be mutually understood by all participants that the pace of research is slower in a PUI environment for many well-known reasons. The student researchers are balancing research with full course loads, and, similarly, the PUI faculty are also teaching five or more courses per year. It should be recognized that it is not the norm for PUI faculty to 'buy out' of teaching. Most PUI labs are at their maximum effectiveness during the summer months, when students are in the lab 40 hours a week without the distractions of classes.

**Talk about the Money**

In the excitement of embarking on a collaboration, it can feel uncomfortable to talk directly about the financial needs of the project, but, ultimately, efforts directed towards financial planning help to improve the chances of successful outcomes of the research.

The difference in research cultures between PUI and R1 environments is perhaps most evident in how they fund science. At a PUI, the research groups do not have to face charges for utilities such as phone lines, IP addresses, printing, etc. Instrumentation at PUIs is operated without user fees, since it is the mission of the PUI to have direct student use of scientific instruments and facilities. In our experience, the majority of PUIs also operate stockrooms that supply a variety of consumables and other lab equipment at no charge, again for promoting the mission of undergraduate research. What will a PUI research group need? Specialized chemicals and equipment, summer stipends for undergraduate students and for the supervising faculty, and travel funds to present undergraduate posters all need to be considered.

Because PUI work often has some (limited) levels of support and can be less expensive overall, the PUI group can often make initial progress with modest initial funding or even with existing resources. Yet remember to ensure that planning considers the costs the PUI group will face for the duration of the work; that will help the PUI group in achieving end-goals of the project. To reiterate a comment in the previous section, any preliminary or exploratory work should be pursued prior to formalizing the collaboration.

It is equally important to recognize that there are very diverse PUI settings that can face serious limitations in resources and in institutional support and recognition of research programs. For example, not all PUIs are able to support summer research programs; if this is the case, one strategy is to work with the R1 partner to host undergraduates over the summer in research internships. The PUI institution may not provide dedicated student/faculty research space, and we have seen colleagues develop agreements to make a teaching laboratory into a dual teaching and research space. Another approach to enhance research capability is to implement course-based research in teaching lab space, such as a project based PChem lab (*40*). An overlooked difference in the financial models at PUIs and R1s is that PUI groups tend to have significantly less access to discretionary funds with which to participate in collaborations. Indeed, an R1 collaborator may not be aware that the PUI group has no access to small pots of discretionary funding to get started. Acquiring an extra computer and some startup chemicals to work on a project may be a much more challenging proposition at a PUI than it is at an R1. We have seen examples in which the R1 institution has donated chemicals, stocks, plasmids, etc. to help overcome this barrier for the PUI group to perform its work.

In the current environment, we see less evidence of PUI and R1 participants writing grants together, although see above on Brent Feske's success. More commonly, funding for the PUI laboratory is achieved through subawards from a grant that belongs to the R1 group or as independent grants written by the PUI faculty member, such as to the NSF (Research at Undergraduate Institutions, RUI) or the NIH (R15- AREA). It is of tremendous value for a PUI faculty member to have her or his own external grant funding. Having such funds is a sign to the scientific world that your ideas for science have been peer reviewed and are worthy of funding. It is important to inquire at each institution about the availability of internal funds to promote PUI/R1 collaborations. It is our experience that administrations on both sides are enthusiastic about initiating PUI/R1 joint research and may be able to provide financial assistance towards such collaborations.

There are other opportunities to obtain money for PUI/R1 collaborations. One program is the NSF EPSCoR and NIH IDeA funding programs, which are available in 24 states and three U.S. territories. NSF set up the Experimental Program to Stimulate Competitive Research in 1979, because of the uneven distribution of federal research funds that flowed to the largest centers. NASA and DEA also have programs, but NIH's Institutional Development Award is the largest of these programs. (See http://www.epscorideafoundation.org/ for more details.) Each state competes for and distributes their monies differently, but most have opportunities for PUI faculty and R1 faculty to collaborate. Often the R1 faculty are looking for collaborators in order to enhance the Broader Impacts that comes from such a collaboration.

One NIH IDeA program is the IDeA Networks of Biomedical Research (INBRE, see https://www.nigms.nih.gov/Research/DRCB/IDeA/Pages/INBRE.aspx), and Furman has received $5.5 million since 2005 through this program.

In many states with INBRE, a single, central R1 institution is the "lead" and the bank, doling out funds to support individual research students that come to their campus in an REU-like capacity, or students working at the PUI on a case-by-case basis where there is direct research collaboration with the R1. South Carolina has three R1s and nearly 30 PUIs, which is different than most of the 27 IDeA/EPSCoR entities, so South Carolina hands out subawards from the USC School of Medicine, and also has a statewide competition for special individual investigator awards. Furman currently has three of the seven available grants in this competition.

Another model in South Carolina is the NSF EPSCoR Track-1 grant, which is thematically centered on advanced materials. John Wheeler of Furman led the writing of this grant, which brought in $20 million while he was running the EPSCoR office. The State then divides the $20 million into subawards for PIs at the different institutions, as well as making separate stimulus funds available to fund smaller grants for faculty at PUIs working with lead faculty at R1s.

If you work in a non-EPSCoR state, there is money available for collaborations from large centers at R1s. NSF and other agencies run large-center grants, such as Science and Technology Centers (STCs), Materials Research and Engineering Centers (MRSECs), Centers for Chemical Innovation (CCIs), and EPSCoR. These are often multi-institutional with plenty of opportunity for PUIs to engage. Greg Springsteen at Furman has collaborated with a CCI in chemical evolution at Georgia Tech for the last decade that has principally contributed to his research funding since 2010, funding students, equipment, supplies, and summer support, and that has allowed him to develop research of high quality published in Nature Communications this year (*41*). One key aspect is that R1 faculty often don't know the PUI talent in their region or even in their city, in some cases. It is up to the PUI faculty to reach out and make those connections at meetings, through invited seminars to the PUI campus, in cold call e-mails, or by whatever opportunity presents itself. R1 faculty are too busy (and do not know how) to look around for that talent at PUIs, but, once it presents itself, they are often thrilled to accept it.

One point we do not think can be overemphasized is that undergraduate research is much less expensive than paying for a graduate student and that federal agencies are so interested in building workforce development and obtaining broader impacts that subawards to PUIs can make R1 applications more competitive. There may also be potential for CAREER applicants at R1s, for example, to engage with local PUI programs to enhance their educational components. As of this writing, NSF still makes available Research Opportunity Awards (ROAs, see https://www.nsf.gov/pubs/2014/nsf14579/nsf14579.htm) that support PUI faculty working at R1 institutions. John Wheeler did this for his first sabbatical, and it was an exceptional experience that greatly benefited his subsequent research program.

### Talk about Dissemination

Discuss in advance the likely assignment of author ordering on manuscripts, particularly on designations of corresponding (a.k.a. communicating) authorship. Many journals permit two corresponding authors. The principal investigators (PIs) from both the PUI and the R1 teams can be designated as a communicating author. Of course, if a PUI or R1 faculty member is not involved in the strategic direction and intellectual understanding of the research, then that person should not be a corresponding author.

PUI researchers may not have been exposed to the complex scoring formulas that are increasingly applied to research faculty at R1 institutions, which may dictate that certain author positions or journal types are preferred and ensure the R1 participants are correctly recognized internally. Asking about scoring formulas is not taboo, and transparent discussions by Rovnyak and coworkers on this subject led to the determination of author ordering in a recent paper that represented norms from both institutions (*7*).

It is recommended that the PI at each institution approves the authors from their institution. That is, the PI from one group would not decide on who earned authorship from the other institution and vice versa. This is important because there can be differing norms for recognizing effort with authorship. For example, PUI groups award authorship to students who perform significant parts of the work and contribute intellectually in other ways but who are not yet at the stage of being able to write the results on their own. In R1 environments, technicians might not appear as authors. Therefore, it is appropriate, in our experience, that each PI applies the norms of their institution when deciding on authorship.

A valued aspect of research at PUIs is that faculty and students tend to participate strongly in conference poster sessions at national and regional ACS meetings and at smaller disciplinary meetings in the area of the PI's research. The PUI group should convey their expectations for sending undergraduate students to conferences to present on the collaborative work. It is extremely important to decide in advance on venues and on which portions of the work can be disclosed at undergraduate poster sessions. Remember to share drafts of posters between groups and to obtain mutual permissions.

Writing manuscripts is a significant effort and some preliminary discussions should take place, detailing which individuals are likely to do the primary writing, figure preparation, and other aspects of the manuscript. Although it is less common for undergraduate students to do the primary writing, many faculty at PUIs work to include the undergraduate students in specific portions of the writing, and this depends on the experience level of the undergraduate. Many of our students have written outlines and are capable of writing up the methods and results sections. We have had seniors with three or more years of research experience and who have already co-authored papers in our groups write the first drafts of papers, but this is a rarer occurrence.

On the subject of publications, discuss whether specific venues are required. Generally, PUIs face less pressure to publish in journals that meet certain thresholds for impact factor, for example, whereas R1 collaborators may need to focus on certain journal titles. So for instance, when GCS was starting out, he often published in the *International Journal of Quantum Chemistry* because that was the journal that was associated with the Sanibel Symposium, where his students presented every year from 1994-2010. He now favors publishing in ACS journals as much as possible, which has had the effect of giving him more visibility in the American chemical community.

A research collaboration may generate intellectual property (IP) and/or inventions, which may be unplanned and even serendipitous. Unexpected discoveries can and do occur. Each investigator can hold internal conversations with respective administration representatives (e.g. sponsored research officers), who will then advise on whether formal agreements between institutions are recommended. DR participates in a collaboration between Bucknell University and Geisinger Research, where the respective institutions have previously agreed on IP and data sharing policies. GCS's collaboration with Albany Medical College eventually resulted in two patents on molecules that inhibited breast cancer in mice and rat model systems (*42*, *43*), and the patent expertise at Albany Medical College was an incredible asset to his group at that time.

## Cruise Control: Doing the Research

At R1s, senior graduate students and postdocs commonly drive the research forward. At PUIs, while these projects are driven by the undergraduate researchers, the professor is usually more directly involved to ensure the progress and timeliness of the collaboration. We have both spent significant time throughout our careers in a hands-on mode to make sure that the data collection is correct or that the results of a simulation have followed the agreed-upon protocol. Students learn through experience, and making mistakes is part of that process, so the learning curve is a bit longer for students at a PUI. Most PUI faculty cannot afford a postdoctoral associate and thus rely on their own expertise to ensure that the research is progressing satisfactorily.

Communication is critical. It is often the case that results from the PUI lab do not come as fast as was originally planned. Of course, adapting time frames for results is common in research, but PIs at PUIs operate under special circumstances that may cause them to miss agreed-upon deadlines for the project. The worst thing a PUI faculty member can do is to fail to respond in a timely fashion to an inquiry from their R1 colleague. If no results have been obtained in the previously-agreed timeline, then the PUI faculty member should acknowledge this, with a minimum of excuses, and provide a more realistic timeline based on the experience gained from missing the previous deadline. Avoid giving the impression that one group will get in touch with the other when the work is completed; instead, concentrate on good communication between groups. All research yields challenges and problems that need to be overcome, and it is critical to recognize the importance of open discussions around the problems that arise on either side of the collaboration.

Most times the R1 lab moves at a faster pace than the PUI lab. This is a natural outcome of the research environment and the faculty reward incentives in these two different types of institutions. If the R1 lab decides they need results more quickly, then they should convey this to the PUI lab before they complete the work in-house or pick up another collaborator. Nothing is more demoralizing than seeing a paper that scoops one's research appear from a collaborator's group; so, R1 faculty should avoid accidentally competing with a PUI collaborator.

Schedule joint meetings, visits, and video interactions. Make sure the collaboration has facetime in person as well as virtually. All of our students learn from these group interactions, as the PIs model what excellent research collaborations look like. All of our students need to learn how to work cooperatively in groups, and the time spent communicating about research processes and results is valuable. When thinking about presentations, remember that both PUI and R1 faculty depend on giving talks, and students make poster presentations, so it is crucial to share slides about the work as it develops. Presentations should always acknowledge the other group's contributions.

Although we reiterate that PUI/R1 collaborations tend to have more carefully defined goals, it is still to be expected that research directions will evolve and change just as in any other research endeavor. Furthermore, the conduct of the research may even lead to the need to consider whether the collaboration should be suspended or ended. Typical scenarios include uncovering results that obviate a goal or discovering challenges that need extraordinary resources that exceed those available in one or both labs. It is difficult in any research endeavor to recognize when a project should be ended, and this is no less true in a PUI/R1 collaboration. When a collaboration ends, remember it is no different than the loss of an internal research project. Do not miss the chance to identify publishable milestones even if the end-result was not achieved. Remember to see this as an opportunity to look for new beginnings, including holding a discussion of how the results and data that were generated up to the point at which the work was terminated will be used by the PUI and R1 teams in future grant proposals. A situation to avoid is adiabatically losing interest by one or both parties; we have seen and experienced this effect, and it can come from either lab. A reasonably diplomatic approach in such situations is to acknowledge that we all are facing extraordinary demands on our time and to ask the partner if they feel the work is the best use of everyone's resources going forward. If one partner is fading out, then they are likely very receptive to conversations about ending the collaboration.

## Road-Tripping: Teachable Moments in R1/PUI Collaborations

### Some Lessons Learned

Our experience and that of many of our colleagues across a wide range of institutions reveal that many R1 and PUI institutions have access to small pots of seed and incentive monies to help start collaborations. Make sure you know what pots of money are available at your institution! To emphasize a point above, it is crucial to inquire of administrators whether money to assist R1/PUI collaboration is available and to advocate for the importance of such support; even asking may itself provide the stimulus for an administration to set up a new initiative to provide such funds.

If you have read this far, then it should be clear to you that *communication* is essential to a successful PUI-R1 collaboration. Do not let gaps in communication develop. It is critical to stay in touch, especially when hitting roadblocks and time delays. Everyone is busy, but everyone is also invested in the success of their research, so make the time to communicate clearly and often.

In our experience, a PUI cannot expect to have a successful collaboration if they only plan on doing research in the summer. Successful collaborations require year-round work, and, even if the pace is slowed during the year, scientific advances do not wait for the summer months. No one likes being scooped. A useful tip for PUI faculty is to do one activity every day of the year that advances one's research agenda.

PUI faculty build their professional recognition and the respect of their peers slowly over time. We have both observed that, as your discipline becomes more aware of your expertise and contributions, developing professional relationships that grow into collaborations will follow naturally. So stick to doing the best science you can, write grant proposals and publish, and you will find the best collaborations for your group.

### Going Off Road: When Collaborations Develop Troubles

Throughout this chapter, we have tried to share anecdotes that speak to specific points, but there are also a number of 'teachable moments' in PUI/R1 collaborations that we think ought to be highlighted. We ask the reader to understand that some of these are sensitive matters, many of which involved colleagues and not us, and that we are not able to provide as much detail as when describing our own experiences.

For example, we are aware of instances in which R1 labs submitted papers about collaborative work without informing the PUI partners. Worse, we have seen the morale-damaging impact that occurs when an R1 lab has published work on which the PUI partner was working, effectively scooping the PUI partner in the joint work. Most stunning of all, we have seen an example in which a PUI partner sent his grant proposal to a funding agency, only to find that the R1 partner had submitted the exact same proposal to another funding agency. Fortunately, this was caught (the PUI partner received his own grant proposal to review and contacted his R1 partner, asking him to withdraw it; he did, and apologized profusely, but one can imagine the strain that this put on that relationship). We are all busy, and it is always tempting to cut corners, but both partners need to respect each other's intellectual property.

We have also seen a situation in which information was shared among R1 partners, but the PUI partners were excluded, leading to a broader point. Communication of results is not a one-way street; successful collaborations are true partnerships and not outreach projects from the R1 institution to the PUI. In general, when we have witnessed collaborations that struggle, it is because communication has broken down, and (unspoken) expectations have not been met. Sometimes expectations have been assumed and not spelled out in advance, and this is something to be avoided in successful collaborations.

## Driving on the Autobahn

Most PUIs offer the opportunity for sabbaticals, and there is nothing like an overseas sabbatical to forge new paths. When GCS was a newly-tenured Associate Professor at Lake Forest College, he had a wonderful sabbatical with Modesto Orozco at the University of Barcelona for 15 months, which was fully funded by the Spanish Ministry of Science and Education. While his three children were learning Spanish and Catalan as they made friends in the local school, he was adding molecular dynamics to his computational expertise in quantum chemistry (*44*, *45*). He learned how European science worked, and his collaboration with Orozco, and Charlie Laughton at the University of Nottingham, was central to his future scientific success.

Successful international collaborations often start with a personal connection and a visit to the host country. Jason Cody, Professor of Chemistry at Lake Forest College, met a postdoctoral fellow from France in the lab where he was working on his Ph.D. Three years later, Cody was an NSF International Postdoctoral Fellow at the Institut des Matériaux in Nantes, France. While working on various projects during his postdoctoral year and subsequent sabbaticals, other international collaborations sprang forth spontaneously from common interests discovered during visits by other international researchers to the lab in France. As many international researchers must publish their work in English, Cody made many connections and learned a wide range of chemistry by offering to polish the English of manuscripts written by other researchers in Nantes (GCS provided the same service in Spain). Although the International Postdoctoral Fellowship is no longer offered, current NSF opportunities are easily found at the Office of International Science and Engineering (OISE). Other avenues for international contacts include the extended professional network of research advisors and mentors (they know who does what and where). Finally, the Fulbright program run by the Institute for International Exchange for the US State Department is designed to facilitate international intellectual exchange for both postdoctoral fellows and established researchers. For PUI researchers, it is very helpful to explain the pace of research at smaller institutions, including the consequences of having only undergraduates work on projects (full-time international researchers might find it hard to imagine how it is frequently necessary to piece together the results from multiple students over multiple summers or academic years into a single publication). As PUI's produce two parallel academic products (publishable results and students ready to go to graduate school or industrial jobs), seldom in a one-to-one ratio, the pace can seem extremely slow for colleagues who spend nearly all of their time doing research.

Professor Karen Almeida at Rhode Island College just finished a year-long sabbatical in Italy, combining a life-long interest in living and working in Italy with a chance to pursue her science. Almeida learned a lot about protein production and biophysical characterizations, and learned how the Italians are masters of doing science with fewer financial resources. The instrumentation and expertise available during this sabbatical were essential to generating corroborating data, allowing her undergraduate research group to publish their findings.

## Finding New Roads: What Successful PUI/R1 Research Accomplishes

From the PUI perspective, we are always focused on our undergraduates' success. When our labs are productive, we obtain grants and publish papers on interesting science, and our students learn how to do science. They win Goldwater Scholarships and Graduate Fellowships. Those that attend graduate school are successful because they already understand the research process, and they are resilient in the face of adversity. When they get stuck, they figure out what they need to know to advance their research. This development of human capital is one of the most important functions of undergraduate research. The addition of a collaboration with an R1 is an additional experience for PUI students to develop as researchers, learn more about the scientific enterprise, and gain additional confidence about their own abilities. It is no surprise that on a per capita basis, PUIs send more students to graduate school than do research universities (*46*). Products should include posters and, in many cases, at least one publication in a peer-reviewed journal.

Similarly, graduate students and postdocs in the R1 team will emerge with a direct experience that PUI/R1 collaboration contributed to advancing research goals, and will be better equipped to seek such interactions in the future. They will also gain a deeper understanding of the unique aspects of PUI career paths and why they attract top scientists (*1*, *2*).

All research answers some of the original questions, and inevitably results in new questions that can be pursued. One example that GCS experienced in his collaboration with Don Landry at Columbia is worth reciting here. The joint project on developing haptens that mimic the transition state for cocaine hydrolysis led Don to pose a question on whether it was possible to calculate the charge state of a potential transition state analog *before* he synthesized it. This question led to an entirely new line of research in GCS's group, four highly cited papers driven by undergraduates Matt Liptak and Annie Toth (more than 125 citations each, with two approaching 400 citations each) (*47–50*), a reinvigoration of this field of research (*51*, *52*), and eventually a book (*53*). Matt is now a chemistry faculty member at the University of Vermont and Annie is a neuroscience researcher at the University of Wyoming. The Rovnyak lab's series of work on NMR sensitivity opened new opportunities, such as working with colleagues in pharma on a two-volume edited book series on frontiers of structure elucidation (*54*, *55*). These efforts in turn have stimulated a broader change in his lab's research direction towards small molecules and natural products that are the subject of some new R1 collaborations being explored at this time. Good communication leads to enthusiasm for more joint work, and science advances. Possibilities for future funding emerge, and our students are the big winners.

## Acknowledgments

We are grateful to the many colleagues and students who have contributed to the rich undergraduate research programs we have enjoyed. We especially thank Jason Cody, Karen Almeida, Brent Feske, Delana Gajdosik-Nivens, Will Lynch, Scott Mateer, Jim Phillips, Lee Silverberg, Greg Springsteen, and John Wheeler for allowing us to profile some of their experiences in this chapter.

## References


1. Rovnyak, D. S.; Shields, G. C. *How Undergraduate Research Drives Science Forward.* Inside Higher Ed. July 7, 2017. https://www.insidehighered.com/views/2017/07/07/undervaluation-role-undergraduate-research-advancement-scientific-knowledge-essay (accessed Feb 10, 2018).
2. Rovnyak, D. S.; Shields, G. C. *Undergraduate research drives science forward.* University Business May 30, 2017. https://www.universitybusiness.com/article/undergraduate-research-drives-science-forward (accessed Feb. 10, 2018).
3. Sherer, E. C.; Turner, G. M.; Shields, G. C. Investigation of the Potential Energy Surface for the First Step in the Alkaline Hydrolysis of Methyl Acetate. *Int. J. Quantum Chem.* **1995**, *S22*, 83–93.
4. Turner, G. M.; Sherer, E. C.; Shields, G. C. A Computationally Efficient Procedure for Modeling the First Step in the Alkaline Hydrolysis of Esters. *Int. J. Quantum Chem.* **1995**, *S22*, 103–112.
5. Sherer, E. C.; Turner, G. M.; Lively, T. N.; Landry, D. W.; Shields, G. C. A Semiempirical Transition State Structure for the First Step in the Alkaline Hydrolysis of Cocaine. Comparison between the Transition State Structure, the Phosphonate Monoester Transition State Analog, and a Newly Designed Thiophosphonate Transition State Analog. *J. Mol. Model.* **1996**, *2*, 62–69.
6. Sherer, E. C.; Yang, G.; Turner, G. M.; Shields, G. C.; Landry, D. W. Comparison of Experimental and Theoretical Structures of a Transition State Analog used for the Induction of Catalytic Antibodies that Destroy Cocaine. *J. Phys. Chem. A* **1997**, *101*, 8526–8529.
7. Miele, M. M.; Irving, B. A.; Wenrich, B. R.; Martin, P. L.; Rovnyak, D. Reproducibility and Stability of Aqueous Metabolite Levels in Extracted Serum by NMR Spectroscopy. *Curr. Metabolomics* **2017**, *5*, 45–54.
8. Shields, G. C. The Benefits of Forming a Consortium for an NSF-MRI Proposal. *CUR Quarterly* **2002**, *23*, 80–81.
9. Rovnyak, D.; Sarcone, M. D.; Jiang, Z. Sensitivity Enhancement for Maximally Resolved Two-Dimensional NMR Spectra by Nonuniform Sampling. *Magn. Reson. Chem.* **2011**, *49*, 483–491.
10. Paramasivam, S.; Suiter, C. L.; Hou, G.; Sun, S.; Palmer, M. R.; Hoch, J. C.; Rovnyak, D.; Polenova, T. Enhanced Sensitivity by Nonuniform Sampling Enables Multidimensional MAS NMR Spectroscopy of Protein Assemblies. *J. Phys. Chem. B* **2012**, *116*, 7416–7427.
11. Suiter, C. L.; Paramasivam, S.; Hou, G.; Sun, S.; Rice, D.; Hoch, J. C.; Rovnyak, D.; Polenova, T. Sensitivity gains, linearity and spectral reproducibility in nonuniformly sampled multidimensional MAS NMR spectra of high dynamic range. *J. Biomol. NMR* **2014**, *59*, 57–73.
12. Palmer, M. R.; Suiter, C. L.; Henry, G. E.; Rovnyak, J.; Hoch, J. C.; Polenova, T.; Rovnyak, D. Sensitivity of Nonuniform Sampling NMR. *J. Phys. Chem. B* **2015**, *119*, 6502–15.
13. Palmer, M. R.; Wenrich, B. R.; Stahlfeld, P.; Rovnyak, D. Performance tuning non-uniform sampling for sensitivity enhancement of signal-limited biological NMR. *J. Biomol. NMR* **2014**, *58*, 303–314.

14. Palmer, M. R.; Gupta, R.; Richard, M.; Suiter, C. L.; Hoch, J. C.; Polenova, T.; Rovnyak, D. Application of nonuniform sampling for sensitivity enhancement of small molecule heteronuclear correlation NMR spectra. In *Modern NMR Approach for the Structure Elucidation of Natural Products*; Martin, G. E., Williams, A. J., Rovnyak, D., Eds.; RSC Publishing: Cambridge, U.K., 2015; Vol 1, pp 93−118.
15. Silverberg, L. J.; Tierney, J.; Cannon, K. C. Research at Predominantly Two-Year Campuses of Penn State. In *The Power and Promise of Early Research*; Murray, D. H., Obare, S. O., Hageman, J. H., Eds.; ACS Symposium Series 1231; American Chemical Society: Washington, DC, 2016; pp 83−118.
16. Shields, G. C. Creating a Comprehensive Summer Undergraduate Research Program Despite Fiscal Challenges. *CUR Quarterly* **2010**, *30*, 20–23.
17. Shields, G. C.; Hewitt, G. J.; North, L. Using Pre-College Research to Promote Student Success and Increase the Number of Science Majors. *CUR Quarterly* **2010**, *31*, 43–47.
18. Shields, G. C. Gajdosik-Nivens, D. A.; Ness, T. Using Early Introduction to Research to Increase STEM Majors: A Tale of Two Colleges, one small highly selective private and one non-selective regional public. In *Educational and Outreach Projects from the Cottrell Scholars Collaborative*; Waterman, R., Feig, A., Eds.; ACS Symposium Series 1248; American Chemical Society: Washington, DC, 2017; pp 107−119.
19. Kirschner, K. N.; Lexa, K. W.; Salisburg, A. M.; Alser, K. A.; Joseph, L.; Andersen, T. T.; Bennett, J. A.; Jacobsen, H. I.; Shields, G. C. Computational Design and Experimental Discovery of an Anti-estrogenic Peptide Derived from Alpha-Fetoprotein. *J. Am. Chem. Soc.* **2007**, *129*, 6263–6258.
20. Lexa, K. W.; Alser, K. A.; Salisburg, A. M.; Ellens, D. J.; Hernandez, L.; Bono, S. J.; Michael, H. C.; Derby, J. R.; Skiba, J. G.; Feldgus, S.; Kirschner, K. N.; Shields, G. C. The Search for Low Energy Conformational Families of Small Peptides: Searching for Active Conformations of Small Peptides in the Absence of a Known Receptor. *Int. J. Quantum Chem.* **2007**, *107*, 3001–3012.
21. Joseph, L. C.; Bennett, J. A.; Kirschner, K. N.; Shields, G. C.; Hughes, J.; Lostritto, N.; Jacobson, H. L.; Andersen, T. T. Antiestrogenic and anticancer activities of peptides derived from the active site of alpha-fetoprotein. *J. Peptide Science.* **2009**, *15*, 319–325.
22. Temelso, B.; Alser, K. A.; Gauthier, A.; Palmer, A. K.; Shields, G. C. Structural Analysis of alpha-Fetoprotein (AFP)-like Peptides with Anti-Breast-Cancer Properties. *J. Phys. Chem. B* **2014**, *118*, 4514–4526.
23. Zhan, C.-G.; Deng, S.-X.; Skiba, J. G.; Hayes, B. A.; Tschampel, S. M.; Shields, G. C.; Landry, D. W. First-principle studies of intermolecular and intramolecular catalysis of protonated cocaine. *J. Comput. Chem.* **2005**, *26*, 980–986.
24. Dunn, M. E.; Pokon, E. K.; Shields, G. C. The ability of the Gaussian-2, Gaussian-3, Complete Basis Set-QB3, and Complete Basis Set-APNO model chemistries to model the geometries of small water clusters. *Int. J. Quantum Chem.* **2004**, *100*, 1065–1070.

25. Day, M. B.; Kirschner, K. N.; Shields, G. C. Pople's Gaussian-3 model chemistry applied to an investigation of $(H_2O)_8$ water clusters. *Int. J. Quantum Chem.* **2005**, *102*, 565–572.
26. Day, M. B.; Kirschner, K. N.; Shields, G. C. Global Search for Minimum Energy $(H_2O)_n$ Clusters, n=3-5. *J. Phys. Chem. A* **2005**, *109*, 6773–6778.
27. Dunn, M. E.; Evans, T. M.; Kirschner, K. N.; Shields, G. C. Prediction of Accurate Anharmonic Experimental Vibrational Frequencies for Water Clusters, $(H_2O)_n$, n=2-5. *J. Phys. Chem. A* **2006**, *110*, 303–309.
28. Morrell, T. E.; Shields, G. C. Atmospheric Implications for Formation of Clusters of Ammonium and 1–10 Water Molecules. *J. Phys. Chem. A* **2010**, *114*, 4266–4271.
29. Shields, R. M.; Temelso, B.; Archer, K. A.; Morrell, T. E.; Shields, G. C. Accurate Predictions of Water Cluster Formation, $(H_2O)_{n=2-10}$. *J. Phys. Chem. A* **2010**, *114*, 11725–11737.
30. Temelso, B.; Morrell, T. E.; Shields, R. M.; Allodi, M. A.; Wood, E. K.; Kirschner, K. N.; Castonguay, T. C.; Archer, K. A.; George C. Shields, G. C. Quantum Mechanical Study of Sulfuric Acid Hydration: Atmospheric Implications. *J. Phys. Chem. A* **2012**, *116*, 2209–2204.
31. Temelso, B.; Archer, K. A.; Shields, G. C. Benchmark Structures and Binding Energies of Small Water Clusters with Anharmonicity Corrections. *J. Phys. Chem. A* **2011**, *115*, 12034–12046.
32. Pérez, C.; Muckle, M. T.; Zaleski, D. P.; Seifert, N. A.; Temelso, B.; Shields, G. C.; Kisiel, Z.; Pate, B. H. Structures of Cage, Prism, and Book Isomers of Water Hexamer from Broadband Rotational Spectroscopy. *Science* **2012**, *336*, 897–901.
33. Pérez, C.; Lobsiger, S.; Seifert, N. A.; Zaleski, D. P.; Temelso, B.; Shields, G. C.; Kisiel, Z.; Pate, B. H. Broadband Fourier Rotational Spectroscopy for Structure Determination: The Water Heptamer. *Chem. Phys. Lett.* **2013**, *571*, 1–15.
34. Pérez, C.; Zaleski, D. P.; Siefiert, N. A.; Temelso, B.; Shields, G. C.; Kisiel, Z.; Pate, B. H. Hydrogen Bond Cooperativity and the Three-Dimensional Structures of Water Nonamers and Decamers. *Angew. Chem., Int. Ed.* **2014**, *53*, 14368–14372.
35. Richardson, J. O.; Pérez, C.; Lobsiger, S.; Reid, A. A.; Temelso, B.; Shields, G. C.; Kisiel, Z.; Wales, D. J.; Pate, B. H.; Althorpe, S. C. Concerted Hydrogen-Bond Breaking by Quantum Tunneling in the Water Hexamer Prism. *Science* **2016**, *351*, 1310–1313.
36. Steber, A. L.; Pérez, C.; Temelso, B.; Shields, G. C.; Rijs, A. M.; Pate, B. H.; Kisiel, Z.; Schnell, M. Capturing the Elusive Water Trimer from the Stepwise Growth of Water on the Surface of a Polycyclic Aromatic Hydrocarbon Acenaphthene. *J. Phys. Chem. Lett.* **2017**, *8*, 5744–5750.
37. Temelso, B.; Klein, K. L.; Mabey, J. W.; Pérez, C.; Pate, B. H.; Kisiel, Z.; Shields, G. C. Exploring the rich potential energy surface of the water undecamer. *J. Chem. Theory Comput.* **2018**, *14*, 1141–1153.
38. Wenrich, B. R.; Sonstrom, R. E.; Gupta, R. A.; Rovnyak, D. Enhanced biosynthetically directed fractionalcarbon-13 enrichment of proteins for backbone NMR assignments. *Protein Expr Purif.* **2015**, *115*, 1–10.

39. Craft, D. L.; Sonstrom, R. E.; Rovnyak, V. G.; Rovnyak, D. Nonuniform sampling by quantiles. *J. Magn. Reson.* **2018**, *288*, 109–121.
40. Shields, G. C. The Physical Chemistry Sequence at Liberal Arts Colleges: The Lake Forest College Approach. *J. Chem. Educ.* **1994**, *71*, 951–953.
41. Springsteen, G.; Yerabolu, J. R.; Nelson, J.; Rhea, C. J.; Krishnamurthy, R. Linked cycles of oxidative decarboxylation of glyoxylate protometabolic analogs of the citric acid cycle. *Nat. Commun.* **2018**, *9*, 91.
42. Andersen, T. T.; Bennett, J. A.; Jacobson, H. I; Shields, G. C.; Kirschner, K. N. U.S. Patent 7598342, 2009.
43. Andersen, T. T.; Bennett, J. A.; Jacobson, H. I; Shields, G. C.; Kirschner, K. N. U.S. Patent 7964701, 2011.
44. Shields, G. C.; Laughton, C. A.; Orozco, M. Molecular Dynamics Simulations of the d(T·A·T)Triple Helix. *J. Am. Chem. Soc.* **1997**, *119*, 7463–7469.
45. Shields, G. C.; Laughton, C. A.; Orozco, M. Molecular Dynamics Simulation of a PNA·DNA·PNA Triple Helix in Aqueous Solution. *J. Am. Chem. Soc.* **1998**, *120*, 5895–5904.
46. Cech, T. R. Science at Liberal Arts Colleges: A Better Education? *Daedalus* **1998**, *128*, 195–216.
47. Toth, A. M.; Liptak, M. D.; Phillips, D. L.; Shields, G. C. Accurate relative $pK_a$ calculations for carboxylic acids using complete basis set and Gaussian-n models combined with continuum solvation methods. *J. Chem. Phys.* **2001**, *114*, 4595–4606.
48. Liptak, M. D.; Shields, G. C. Accurate $pK_a$ Calculations for Carboxylic Acids Using Complete Basis Set and Gaussian-n Models Combined with CPCM Continuum Solvation Methods. *J. Am. Chem. Soc.* **2001**, *123*, 7314–7319.
49. Liptak, M. D.; Shields, G. C. Experimentation with different thermodynamic cycles used for $pK_a$ calculations on carboxylic acids using Complete Basis Set and Gaussian-n Models combined with CPCM Continuum Solvation Methods. *Int. J. Quantum Chem.* **2001**, *85*, 727–741.
50. Liptak, M. D.; Gross, K. C.; Seybold, P. G.; Feldgus, S.; Shields, G. C. Absolute $pK_a$ Determinations for Substituted Phenols. *J. Am. Chem. Soc.* **2002**, *124*, 6421–6427.
51. Alongi, K. A.; Shields, G. C. Theoretical Calculations of Acid Dissociation Constants: A Review Article. *Annu. Rep. Comput. Chem.* **2010**, *6*, 113–138. Wheeler, R. A., Ed., Elsevier.
52. Seybold, P. G.; Shields, G. C. Computational estimation of $pK_a$ values. *WIRES Comp. Mol. Sci.* **2015**, *5*, 290–297.
53. Shields, G. C.; Seybold, P. G. *Computational approaches for the prediction of* $pK_a$ *values*; CRC Press: Boca Raton, FL, 2014; pp 1−176.
54. *Modern NMR Approaches for the Structure Elucidation of Natural Products*; Williams, A. J., Martin, G. E., Rovnyak, D., Eds.; RSC Publishing: Cambridge, U.K., 2017; Vol. 2, Data Acquisition and Applications to Compound Classes; pp 1−516.
55. *Modern NMR Approaches for the Structure Elucidation of Natural Products*; Williams, A. J., Martin, G. E., Rovnyak, D., Eds.; RSC Publishing: Cambridge, U.K., 2016; Vol. 1, Instrumentation and Software; pp 1−329.

## Chapter 12

# Case Study: A Roadmap to Successful Collaborations Between Primarily Undergraduate Institutions and Research Institutions

**John D'Angelo**

**Division of Chemistry, Alfred University, Alfred, New York 14802, United States**
***E-mail: dangelo@alfred.edu.**

When it comes to publication time, collaborations present a layer of complication. Navigating this is not always easy especially when the collaboration is between radically different types of institutions. Herein are cases that explore how coauthorship relates with respect to all authors seeing a paper before publication.

## Imagine This…

Jose Garcia is a faculty member at a Primarily Undergraduate Institution (PUI), a small liberal arts college without a graduate program in chemistry. He directs undergraduate research and has a project in which his students are pursuing small-molecule inhibitors of *e-coli.* Jose, earlier in his career as a graduate student, collaborated with a lab at a major medical research university that had precisely the expertise needed to test his current lab's compounds. Jose's institution could never test the compounds due to a lack of adequate facilities to safely work with such pathogens. After a few years, his lab has created enough compounds to test, and Jose sends them out. A few months go by, and Jose's star student, Jennifer Janey, comes into his office, visibly upset. Jose asks her what is wrong, and she responds by saying, "I thought we were a team! I really wanted to be a part of writing this paper, but, not only did you not let me, you didn't even tell me when it got accepted."

Jose, now as confused as Jennifer is angry, responds calmly "I have no idea what you're talking about. I haven't even seen the cellular assay results yet; how can anything be published?" Jennifer then hands him the preprint of their work, and he finds that, to his shock and disappointment, the collaborating lab has published their work, adding Jose and Jennifer to the author list without ever including them in the writing or discussion of the results. He calls the lab and demands an explanation, and they respond, saying that:

- The work was finished over the summer, they assumed that Jose and Jennifer were not around, and they did not want to delay the publication;
- They assumed that, since the article was not really about the synthesis, Jose and Jennifer's work would just be included in the supplemental information, leaving little for them to contribute to the actual paper, as it is outside their expertise; and
- They assumed that having a researcher from the more prestigious lab as the corresponding author would make the peer review process far easier.

After hanging up, Jose recounts the exchange to Jennifer and indicates to her that they will no longer collaborate with that group.

*Questions to ponder:*

- Should Jose and Jennifer just be grateful that so much work was done for them, or do they have a legitimate complaint against the collaborator?
- What could the journal have done better to verify the consent of all authors since, after all, this was a multi-institution collaboration?
- Is it acceptable to exclude coauthors from the writing and editing of an article if much of the paper is outside of their areas of expertise?
- What problems may arise for Jose and Jennifer if any aspect of the paper is ever questioned? Why is it important for all authors to approve the publication?
- Should the prestige of an author matter when it comes to peer review?

Although cases like this hypothetical one could be avoided if publishers required a signature (electronic or paper) from all authors, weaknesses in the process have been exploited and thus such measures may not fully help situations like that described in this hypothetical case.

For instance, sometimes authors can be added fraudulently to help a paper get approved. It is a dirty little (not) secret in peer review that a distinguished scientist's work is reviewed less critically than that of a newcomer. In truth, this is one of the weak spots in the peer review process and one piece of evidence that can be cited in favor of a double-blind peer review process. It is also something that nefarious authors have done to try to get ahead.

Such was the case when a group of researchers at a few universities in Iran added a prominent Dutch economist to an article for a management science journal. The authors were not even savvy about their deceit, using a non-institutional email address as the contact information for the Dutch economist, whom they listed as the corresponding author. Such odd behavior alarmed the editors of the journal, who took the liberty of contacting the prominent researcher at his normal institutional email address. He informed them that he had nothing to do with this paper and that this was not the first time that something like this had happened. Suspecting what was going on, the editors authored fake reviews and sent them back to the authors, who immediately revised and returned the manuscript, *sans* the prominent Dutch researcher. When asked about the author change, the authors claimed that, after the revision, he no longer wanted to be an author of the paper. To find out what the editors did, consult the link below.

*Questions to Ponder*

- What should the editors do?
- Since the final submission did not include the prominent researcher, was there any real foul?
- Did the editors behave unethically in setting such a trap? Would it perhaps have been better to directly confront the authors?

## Source

McCook, A. Retraction Watch. *A new way to fake authorship: Submit under a prominent name, then say it was a mistake.* November 28, 2016. http://retractionwatch.com/2016/11/28/new-way-fake-authorship-submit-prominent-name-say-mistake/ (accessed January 19, 2018).

## Chapter 13

# Authorship in Undergraduate Research Partnerships: A Really Bad Tango Between Undergraduate Protégés and Graduate Student Mentors While Waiting for Professor Godot

**Amy Andes and Patricia Ann Mabrouk***

**Department of Chemistry and Chemical Biology,
Northeastern University,
Boston, Massachusetts 02115, United States
*E-mail: p.mabrouk@neu.edu.**

There is a considerable disconnect regarding scientific authorship, including its requirements and responsibilities, that troubles many STEM disciplines. Given the importance of authorship in STEM research and the emphasis placed on the early immersion of undergraduates in authentic research experiences, surprisingly little is known about undergraduate students' experience with authorship. Graduate students often act as surrogate research mentors in undergraduate research experiences at many research universities. In this case study, we used a grounded theory approach to probe the experiences of three graduate student mentor-undergraduate student protégé dyads working in in the same laboratory at a research university to learn about their knowledge and experiences with authorship. The diversity of the perspectives of the graduate student mentors and their undergraduate protégés offer evidence that everyone working in the same lab may not agree on the definition of authorship, its requirements, and its responsibilities. Graduate and undergraduate students struggled to define authorship and subsequently were unsure of their role, if any, in authorship decision-making. We found that both graduate and undergraduate students were reluctant to discuss their questions with their mentor or faculty advisor. These findings suggest the need for faculty to take a more active lead in teaching graduate and undergraduate research students about the peculiar requirements and responsibilities of authorship that may exist in their laboratory. More research is required to better understand undergraduate research partnerships so that appropriate interventions can be designed and put into place.

Authorship is widely used to assign credit to those involved in academic and research ventures that result in scholarly publication and therefore is of great importance in science, technology, engineering, and mathematics (STEM). The number of publications appearing on a researcher's biosketch is considered to be evidence of past productivity and indicators of likely success when evaluating grant proposals for federal funding. Research productivity is often assessed based on one's position in the byline, the number of one's peer-reviewed publications, and the number of citations one's papers receive. In academia, h-index (*1*), which attempts to quantitatively capture both a researcher's productivity and citation impact, is becoming increasingly popular when evaluating a faculty member's suitability for tenure and professional advancement.

The nature of research, the methods used, and the modes whereby findings are communicated to the greater community of practice vary widely in STEM; so, it is not surprising that the standards for authorship of scientific publications also vary widely not only between disciplines (*2–5*) but even within a single field, such as chemistry (*6*). For example, The Royal Society of Chemistry (*7*) states that, at a minimum, authors must be able to take ownership of a particular section of the manuscript and accept public responsibility for their work. The American Chemical Society (*8*) says that authors should have made significant scientific contributions to the work and accept responsibility and accountability for their findings. Interdisciplinary, global standards such as those of Committee on Publication Ethics (COPE) and the International Committee of Medical Journal Editors (ICMJE) (*9*, *10*) and methodologies such as the Contributor Roles Taxonomy (CRediT) and the authorship matrix (*11*, *12*) have been developed in recent years by scientists and publishers to open a conversation about authorship that recognizes not only the credit-bearing nature of authorship but also its ethical responsibilities (*13*).

Much attention has been paid in particular to the issue of authorship hierarchy and the development of processes to ensure that authors on multi-author studies receive appropriate credit for their contributions (*11–14*); however, there is no norm for establishing authorship order. In one study, in which STEM faculty with graduate training from 15 different institutions were interviewed, faculty were discovered to use a wide array of approaches to determining authorship hierarchy, including listing co-authors alphabetically, ordering them based on the amount of work they contributed, and ordering them based on the faculty's perception of their co-authors relative contributions (*15*).

Given its importance in STEM research, it should not be surprising that authorship is a frequently cited concern for STEM faculty, postdoctoral students, and graduate student researchers working collaboratively (*16*, *17*). Recognizing the challenges, the American Psychological Association Student Science Council published a document for graduate students in 2006, outlining the criteria for determining authorship credit and order (*18*). The document provides useful links and resources for graduate students to help them recognize oft-used criteria and to aid them in authorship discussions and disagreements with their faculty advisors.

Given the lack of disciplinary norms and the variety of approaches that faculty use in authorship decision-making, it is not surprising that there is evidence that postdoctoral fellows, graduate students and undergraduate students may not fully appreciate the implications of authorship (*19–24*). In fact, it is not even clear whether education in issues related to the responsible conduct of research (RCR), including authorship, is effective in changing attitudes or improving ethical decision-making (*25–30*).

In the case of undergraduate researchers participating in research ethics training, evidence suggests that, while many undergraduates can define fundamental research ethics concepts, including authorship, they are unable to apply this knowledge to their research. This inability appears to be due, at least in part, to their lack of knowledge of the ethical standards and conventions in their laboratories (*31*, *32*).

Many studies of authorship in undergraduate student-faculty collaborations cite abuse of power as a common theme (*4*). This issue is also frequently mentioned when graduate students act as research mentors to undergraduates (*33*). Given the complexity of the authorship landscape and the potential for problems, it is not surprising that some caring faculty have sought to create thoughtful guidelines for author credit and order when faculty are working collaboratively with undergraduate (*34–36*) and graduate researchers (*37*).

Given the importance of authorship, we were surprised how little is known about authorship in undergraduate research partnerships. For this reason, we have sought, over the past three years, to understand the timing and process involved when negotiating authorship of papers that result from undergraduate research projects. Because we wanted to understand the process from the vantage point of the participants, we chose to use a qualitative research methodology: specifically, grounded theory (*38–41*).

Grounded theory was an appropriate choice of methodology for our work for several reasons. First, it is a well-respected qualitative research method used widely in the social sciences. Second, grounded theory is often used when one seeks to develop an understanding of how different participants experience a mutually negotiated process. In our case, we seek to understand the behavioral processes underlying the negotiation of authorship in order to explain the individuals' ethical decision-making processes. In the long term, we seek to develop strategies to better inform all parties involved in the authorship decision-making process in STEM undergraduate research (UR) partnerships.

In the process of recruiting participants to our study (*42*), one graduate student mentor of an undergraduate researcher recruited two more graduate students and their undergraduate research protégés from the same laboratory (snowball sampling). This provided us with a sample consisting of three graduate student mentor-undergraduate student protégé dyads, all working in the same research laboratory. We thought this was an excellent opportunity to investigate the phenomenon of authorship within a single research group, which could be regarded as a single community of practice (*43*). A number of researchers (*44–46*) have successfully argued that research groups operate as "mini-organizations" at research universities and that each has its own unique organizational culture and social and psychological norms (*44–46*). We specifically sought to learn whether students working in the same research group experienced issues related authorship similarly or differently; and, if their experiences were different, how they varied. Since graduate students often serve as mentors to undergraduates engaged in UR experiences at research universities, we wanted to know whether the graduate students informed the understanding of undergraduates regarding authorship, and, if so, how they did.

We offer the present study as a starting point for future studies of authorship in undergraduate research partnerships at research universities. Our results are intended to 1) encourage healthy discussions of authorship among faculty members and between faculty and their graduate and undergraduate student researchers, in hopes that they will lead to the identification of authorship best-practices and 2) inspire others to conduct larger-scale studies into this fascinating phenomenon.

This study provides the first glimpse at authorship experiences from the perspective of three graduate student mentor-undergraduate student protégé dyads within a single research group. The students were engaged in structural biology research in a chemistry and chemical biology department at a large, private, research-intensive university located on the East Coast of the United States.

## Context of the Study

The context of this work is an academic undergraduate research experience in the chemical sciences at a research university. At a research university, undergraduate research is often an educational experience in which an undergraduate works under the guidance of a graduate student in a faculty member's research laboratory on a project assigned and designed by the faculty member. The graduate student may or may not have prior experience supervising undergraduates. Often the graduate student has not received any prior formal or even informal mentor training. The use of graduate students as mentors and their training are issues of current interest (*45, 47–51*).

In this study, all of the undergraduates were working in the same research group under the mentorship of a graduate student assigned to them. Their faculty advisor is known to have a strong interest and commitment to the full participation of undergraduates in research. The graduate students serving as mentors were responsible for providing direct research supervision of their undergraduates. Only one of the graduate students in this study had participated in a formal mentor training program (*52*). Each of the undergraduates had his or her own research project. All the projects that these students pursued focused on research in the same discipline, chemical biology, and all were focused on gaining a better understanding of how a specific class of proteins works. The home department in this study is somewhat unique in that all doctoral students are required to complete a one-semester research ethics course (*53*) that is typically taken during the first year of graduate study.

## Research Methods

This study was approved by the Institutional Review Board at the outset of the study. Signed consent was used.

### Participants

As stated above, the three graduate student-undergraduate research student dyads all worked in the same laboratory at a private research university. To preserve the anonymity of the participants, we will henceforth identify the graduate student mentors as graduate student 1 (G 1), graduate student 2 (G 2), and graduate student 3 (G 3) and their respective protégés will be referred to as undergraduate 1 (UG 1), undergraduate 2 (UG 2), and undergraduate 3 (UG 3). Figure 1 summarizes the critical demographic information provided by the six participants.

All three dyads were same gender-pairings; dyads 1 and 2 were male-male dyads, and dyad 3 was a female-female dyad. Several studies have shown that gender and race can impact both the nature of the mentoring relationship as well as participants' perceptions of their relationship (*54–58*). Male mentors appear to be more likely to provide their protégés with career mentoring while female mentors are more likely to provide their protégés with psychosocial support (*54*, *57*). Female protégés are more likely to see their mentors as role models (*56*, *57*). Protégés from the same race as their mentor appear to receive higher levels of psychosocial mentoring than protégés in cross-race dyads (*55*, *59*). This association raises the question of whether same-gender pairings might affect undergraduate students' receipt of mentoring on authorship. Due to the small number of participants in the present study, this is not a question that we will be able to answer here, but it is one that would merit further study.

### Participant Recruitment

We emailed graduate students acting as mentors to undergraduate research students to solicit their and their protégés' participation. Our email explained the purpose of our study and that we were looking to recruit mentor-protégé teams. All participants were offered a cash honorarium of $50 for their participation in a private interview that we anticipated would last one hour. If graduate student mentors responded to our inquiry and expressed interest in participating in the study, we asked them for permission to contact their undergraduate protégés to find out if their protégés would also be interested in participating. We then reached out to the undergraduate researchers independently to determine if they were interested and willing to participate.

| | Dyad 1 | Dyad 2 | Dyad 3 |
|---|---|---|---|
| Graduate | - Black male<br>- Third year PhD student<br>- Participated in UR<br>- 1 paper as UG and 1 paper as 2nd author under revision<br>- Mentored 3 UGs as an UG and 2 UGs as Grad student | - Hispanic male<br>- Fifth year PhD student<br>- Participated in UR<br>- 2nd author on 1 publication as Grad student<br>- Mentored 2 UGs previously<br>- Plans to do industrial postdoc | -Hispanic female<br>- Third year PhD student<br>- Participated in UR<br>- Worked on unpublished paper as UG<br>- First time mentoring an UG<br>- Plans to teach or go into administration in academics |
| Undergraduate | - White male<br>- Sophomore undergraduate<br>- Chemistry major<br>- First UR experience<br>- Plans to work | -White male<br>- Graduating senior<br>- Biochemistry major<br>- First UR experience<br>- Interested in graduate school | -Black female<br>- Graduating senior<br>- Biochemistry major<br>- Nontraditional student<br>- Transfer student<br>- First UR experience<br>- Plans to work |

*Figure 1. Demographic characteristics of the three graduate student mentor-undergraduate student protégé dyads, who participated in this study, from the same research group.*

One of the graduate student participants spoke to several peers working in the same laboratory about our study. The conversations that ensued between the graduate students led to the further recruitment of three graduate students and their undergraduate protégés from the same laboratory, a technique known as snowball sampling. At the time of the study, one postdoctoral fellow, six graduate students, and three undergraduate students were members of this research group, meaning that we had successfully recruited and interviewed all eligible individuals in this laboratory. We also intended to interview the faculty advisor, but the faculty member was unable to participate, citing a "very intense schedule." This study focuses on the interviews we had with the three graduate student mentor-undergraduate student protégé teams we recruited from this research group.

### Data Source

Two separate scripts were used to guide our interviews with undergraduate students and their graduate student mentors (Appendices A and B). The scripts were written in such a way as to obtain complementary information from the graduate students and their undergraduate protégés. In each case, the discussion was structured around a finite set (*ca.* ten) of thoughtfully designed questions aimed at learning about the participant's background and relevant experiences pertaining to the undergraduate and his or her research, the participant's training in research ethics and authorship, and the participant's understanding of authorship in general and the group's authorship decision-making practices.

### Data Collection

All participants were interviewed separately, using a semi-structured protocol and at a time chosen by the participant. The interviews were held in a location on campus but away from the participants' research laboratory. The interview sessions varied in length but were, on average, thirty-minutes. A digital tape recorder was used to record the interviews so that we could focus entirely on our conversation with the participants, rather than worrying about capturing the conversations on paper. Interviews were later transcribed verbatim and anonymized by one author, after which they were checked for accuracy against the interview recording by the other author. Finally, the transcripts were imported into NVIVO v. 11 (qualitative analysis software) for coding and analysis.

### Data Analysis

An ongoing process of identifying themes, patterns, and ideas was used to code the interview data. In this process, no *a priori* themes or codes were used. Codes were derived organically by the original investigators, each working independently to code each interview (internal validity). The researchers then met to discuss and reconcile coding differences where they existed. As new ideas and patterns emerged, we re-examined prior interviews to see whether they also contained these new codes. This iterative process was continued until no new codes could be identified in the data (constant comparative analysis). Codes were then placed in categories, relationships between categories were compared, and the categories were refined until theoretical saturation had been achieved. The interview data were analyzed in pairs (mentor/protégé) in order to see how well the protégés' perceptions compared with those of their graduate student mentors.

## Results

All three of the graduate students in this study had completed a formal one-semester course in research ethics. Few doctoral programs in chemistry require a formal course in research ethics, so we found it notable that G 3 had completed a course in research ethics at the institution at which she had earned her MS degree prior to taking the research ethics course required at her current institution. None of the students stated that they had received any research ethics training, formal or informal, in their laboratory.

Two of the three undergraduates (UG 1 and UG 3) indicated that they had not participated in any prior training in research ethics. None of the undergraduates were aware of having received any formal or informal training in research ethics in their current research group. The failure of undergraduate researchers to receive RCR training is not peculiar to this laboratory. In fact, none of the undergraduate researchers interviewed in our ongoing STEM study reported having participated in any formal RCR training (*42*). Those faculty and students stated that ethics was only ever discussed informally and usually only when specific issues arose in the process of conducting their research. UG 2, a senior who had participated in two industrial co-op internships before participating in undergraduate research, was uncertain as to whether or not he had received any training. UG 2 thought he might have received training while on co-op and said: **"Probably with [company] I think. They just do, they do a lot of trainings just to cover everything as a co-op and make sure I'm not doing anything illegal."** After the interviewer read this student a list of essential research ethics concepts that might have been part of any RCR training he had received, UG 2 stated: "**Not necessarily authorship, but like confidentiality and conflict of interest and stuff like that. Yeah."**

It seems reasonable that corporate ethics training would focus on those issues of prime importance to the company and therefore encompass issues including confidentiality and conflict of interest. For example, global giant Lockheed Martin's "Voicing Our Values" ethics training program (*60*) includes video case studies on intellectual property, conflict of interest, safety, sexual harassment, and diversity. Since the publication of research findings is not a high priority for most companies, it seems quite reasonable to expect that corporate ethics training might not address issues such as publication authorship.

Two of the graduate students (G 1 and G 2) interviewed had published papers, G 1 as an undergraduate and G 2 as a graduate student. Only one graduate student (G 1) indicated that he had held an explicit conversation about authorship with his research advisor. The third graduate student (G 3), who had no prior authorship experience, said "we haven't gotten to that" suggesting that she believes a conversation will happen at some point in the future.

None of the undergraduates reported having received any formal training or having participated in any discussions with their faculty advisor about authorship.

Among the graduate and undergraduate students alike, there was general confusion and disagreement concerning the definition and requirements for authorship. It was unclear whether the lab uses a definition of authorship in which all members must fulfull the same requirements or whether the lab's requirements for authorship vary based on the level of the researcher's education. Unfortunately, we did not have the opportunity to interview the faculty mentor in this study so we do not know how s/he defines authorship or whether s/he believes that there are different standards for authorship for graduate students and undergraduates.

In this study, the views of the graduate students, all of whom had received formal research ethics training and were working in the same lab, varied widely. G 2 said there was one definition of authorship for everyone and that authorship required the individual to have made a material contribution to the manuscript, in the form of results. G 1 and G 3, on the other hand, stated that the requirements for authorship are different for graduate students and undergraduates. G 1 stated that graduate students must write part of the manuscript and that undergraduates must participate "greatly" in the paper but do not have to write in order to become authors. G 3 felt that the undergraduates either had to write the paper or contribute enough data for a figure in the paper. In other words, despite the fact that all three graduate students were working in the same laboratory, they all saw the requirements for authorship quite differently.

In our conversations with both the graduate and undergraduate students, we saw that everyone agreed that authors needed to contribute in some way to the paper. However, each student's idea of what constituted the necessary contributions differed widely. For some, the quantity of data and analysis was the defining attribute, while for others the nature of the contribution was critical. Students focused on time, effort, and even commitment in determining whether a "contribution" rises to the level of authorship. UG 1 felt that this contribution had to involve some intellectual thought. He said: **"Something that requires at least some level of maybe thought or planning or creativity. You know beyond just, 'Pipet this into here' and then you know just do that a thousand times."**

UG 1 stated that preparing a series of buffers would not meet the minimum requirements for authorship. However, UG 3, working in the same lab, felt that this activity might merit authorship. She said:

> **"I don't know. I have no idea. But I assume that it means that you contributed in some way to the research. I don't know to what extent. Maybe if you're just the person, I don't know, that makes the buffer. Maybe not. I don't know. I'm not sure. But that's all; I really don't know much."**

For the graduate students, the preparation of buffers was universally viewed as being material. G 1 felt that both material contribution (conducting experiments tied to the paper), and non-material contribution ("active participation") were required to meet the standard. G 2 and G 3 felt that generating data for a figure in the paper was adequate. In short, everyone in the research group whom we interviewed had a different opinion as to what constituted authorship, and the views of the graduate student mentors were markedly different from those of their protégés.

In our conversations with graduate and undergraduate students, everyone spoke of authorship as a credit-bearing opportunity. No one implied that he or she recognized that authorship came with ethical and legal obligations and responsibilities. This likely reflects the students' limited knowledge of and experience with authorship to date. Nonetheless, this represents a clear area of educational need for both undergraduate and graduate student researchers.

In our conversations, we saw that a lack of training and information could lead to misconceptions. UG 1, having no prior undergraduate research, reached several erroneous conclusions about authorship based on a presentation made by a graduate student at a recent group meeting. This undergraduate concluded that acknowledgments are a form of authorship, that everyone in a laboratory gets their name "somewhere" on the publications coming out of the lab in which he works, and that authorship is based on "participation." He said:

> **"There, there was a group meeting I sat in on. There was a, a presentation done by one of the graduate students I believe about some of the research that he did, and you know a paper that he was planning on writing. And in it, it listed pretty much everyone in the lab. You know just sort of at the least acknowledgments. So it seems like at the very least my name would be on there somewhere just for having sort of participated at some level in the research, but for actually having some sort of level of authorship on it, it would require a lot more participation in the research."**

UG 1's failure to appreciate the differences between attribution in the by-line as opposed to being mentioned in the "Acknowledgements" section of a paper could, if not addressed, lead to bitterness, loss of credit, etc. with his undergraduate research experience in the long term.

We found that the students' practical experience with authorship could vary widely even though they are working in the same laboratory. UG 3 was unaware of whether any of the other students or graduate students in her laboratory, including her mentor, had published papers or were working on manuscripts. We also learned that students would not ask about authorship if they did not know that the opportunity existed or if they did not understand the concept of authorship. As UG 3 said:

> **"No. It's kind of the same reason I never asked about I don't know, anything else I don't know about. You know what I mean. I don't even know enough to start the dialogue about it. You know, I never even considered it enough. Yeah, other than the fact that it just sounds like a cool thing that I heard about that time I was here for orientation, three and a half years ago, I never really thought about**

**it. And then that job criteria that I saw, but I never really considered it. I have no idea how you go about it so."**

None of the graduate students interviewed knew what decision-making role they might have in determining the authorship of their undergraduate protégés, and none of them had held an explicit conversation about this with their advisor. All three graduate students surmised that they might have some say in whether or not their student would be an author, but they were unclear what their role might be. Given the general confusion as to the requirements for authorship in their lab, this was not surprising. As G 1 said:

> **"Um so that student is directly working with me, so when there comes time to publish that paper, um I imagine, this hasn't been explicitly said, but I imagine I have the opportunity to go to my advisor and say, 'Hey, such student did this much work on the paper. I believe he or she should be second author.' Whatever the authorship is."**

All three of the undergraduates in this study expressed an interest in being authors on a publication resulting from their undergraduate research, but none of them knew what the requirements for authorship were in their research group. We asked them why they had not asked what the requirements for authorship were. Each of the students expressed reluctance to ask because they felt they were too new or didn't know enough. UG 1 explained that he felt he was "very new" to the lab. UG 2 said he had not asked because he wanted to be "comfortable" and know what he was talking about:

> **"I guess I'm still waiting for me personally to be comfortable with the uh information that I'm working with. I'd like to really know what I'm talking about before I say like, "Hey like, let me start working on something like that." You know, I don't want to just ask and then be completely foolish and not really know what I'm doing. I'm someone who really like; I'm pretty careful. Um, I overthink things. So I like to know that I really know what's going on before I make a move."**

UG 3 said she had not asked because she just "didn't know what she didn't know", and so, all three undergraduate students working in this lab are waiting…waiting for Professor Godot.

Moreover, we have no evidence that undergraduate students participating in this study ever did have a conversation with their graduate student mentor or faculty advisor. At the end of the semester, we contacted all three undergraduates to find out whether or not they had spoken with either their graduate student mentor or their advisor to clarify their understanding of authorship. Graduating seniors UG1 and UG 3 did not respond to our enquiry. UG 2, who was preparing to leave the laboratory for a study abroad experience, told us: **"Honestly I totally forgot to ask about authorship with my advisor."**

We therefore have no evidence that these students' understanding of authorship changed after our interviews.

Why does all this matter? The apparent differences in mentor and protégé views on authorship matter because the potential for authorship problems is compounded if the graduate students are unaware of their role in the decision-making process and do not know the norms for authorship in their research laboratory. None of the graduate students from this laboratory with whom we spoke clearly understood their role, if they had one, in the decision-making process for determining authorship for their undergraduate protégés. Also, none of the undergraduate students had any prior knowledge of or experience with authorship. Moreover, as we have shown, everyone in the group with whom we spoke had a different understanding of the criteria for authorship.

The graduate students' failure to speak up and their advisor's failure to explicitly discuss authorship have the potential to lead to longstanding misunderstandings and even hurt feelings for themselves and their protégés. Indeed, G 2 disclosed that he was unhappy about his place in the authorship hierarchy of the one publication he has co-authored to date. However, he felt uncomfortable speaking up and having a direct conversation with his faculty advisor about his place in the by-line. G 3, who was not yet an author, felt that determining authorship is clear-cut, based on her limited observations to date:

> **"...But for the most part, authorship is usually pretty clear I think. In my opinion...usually, there is either one or two authors on each paper, and there is usually someone who is clearly leading up the project. Or it's just a collaboration with another lab, and, you know, we just put, you know, one person our lab is working on that paper. So, they work with them to figure out authorship."**

G 3 appears to be creating an inferential model that may not be transferable to future research and publication opportunities.

## Discussion

Today's student is tomorrow's faculty mentor or principal industrial scientist and group leader. We fail as educators if we do not teach our students core RCR concepts, such as authorship. In the process, we have the opportunity to teach our students how to be effective leaders and managers on topics that are sometimes challenging. Our unwillingness to embrace this challenge has the potential to perpetuate misunderstandings and lead to dissatisfaction that could result in the loss of critical STEM talent.

What is striking to us in this set of interviews is both the heterogeneity and at the same time the homogeneity of the experiences and views expressed by the graduate and undergraduate students. The heterogeneity of views about authorship in dyads and between dyads points to the fact that, while the students may be discussing research methodology, data, etc. with each other, they are not discussing authorship. The graduate students all took a course in research ethics; yet each of them, though they are all members of the same research team, has a different understanding of authorship. The undergraduates are ignorant on this subject, but they will not ask for guidance because they do not know what they do not know. What little everyone does know appears to have been learned through personal experience and through informal exchanges with their peers. Several other researchers have made similar observations: students learn their groups' values through informal interactions with their peers (*15*, *46*, *61*).

It is interesting that we do not seem to be applying the lessons learned from developing effective teaching methods to the ways in which we lead our research teams. In classroom teaching, assessment of student learning drives our work. Are our operational practices in our research groups genuinely aligned with our goals? If so, either we do not value what we espouse or what we are doing isn't working. In either case, the available evidence supports the need for a significant change.

## Limitations

This investigation focused on the phenomenon of authorship in undergraduate research partnerships at one private research university. Our work was based on interviews with three graduate student-undergraduate student dyads engaged in a single type of research in a single academic discipline, specifically, chemistry. As the faculty research advisor declined to participate in our study, we do not have any direct information about the faculty member's definition of authorship or the criteria that he or she uses when determining authorship. The graduate students in this study self-selected for participation. Their action may reflect some interest in and therefore bias related to the topic of the study. Due to the small size of the research group and, therefore, the limited number of available research subjects, there may be other voices and experiences that we did not capture in our study. It is also important to keep in mind that we spoke with these students at one point in time and therefore do not know whether or how these students' understanding of authorship may have changed over time. For all of these reasons, the results from this study should be regarded as a starting point in the identification of the factors and relationships involved in authorship. We intend that they should serve as a foundation for broader studies in a wide array of contexts and not be viewed as generalizable.

## Conclusions

Our results suggest that knowledge of and experience with authorship can take a variety of forms, which can range quite widely. It is clear that additional research on this topic is needed. This subject seems particularly relevant today, given the increasing use of graduate students as surrogate mentors for undergraduates at research universities. While we agree that there are many positive outcomes associated with the use of graduate students as research mentors to undergraduate researchers, the results of our study suggest that a more hands-on approach by the supervising faculty member may be warranted. Active participation by the faculty mentor is needed to ensure that all students participating in research receive training and know what they need to do to earn authorship on papers published by their research group. Ideally, faculty should educate themselves regarding the established definitions and criteria used in their disciplines. At a minimum, faculty should be able to articulate their working definitions and the criteria they use in making decisions about student authorship and discuss these openly with their research teams.

In the present study, the voice of the faculty member is absent. If we borrow from our ongoing study (*42*), faculty frequently cited their workload as a factor contributing to their failure to provide their research students with explicit training on RCR and authorship.

We would like to share several ideas for possible mechanisms of intervention.

## Checklists

First, the lowly checklist has proven extremely useful in some fields including aviation and medicine (*62–68*). Creation of a checklist on crucial RCR issues that could be distributed widely to faculty, graduate students, and undergraduate researchers could be used to jumpstart the needed conversations on critical issues, including authorship, that must take place within the research group. In the highly hierarchical discipline of medicine, the use of checklists has been shown to be beneficial in identifying cultural barriers and empowering all members of a surgical team to act (*69*).

## Peer-Mentoring

Changing the culture of an organization or a discipline is obviously challenging. Educating faculty is one approach. The College of Medicine at Hallym University in Korea developed a one-day, team-based learning course on publication ethics for new faculty supervising graduate students and reported observing a positive change in attitude among the faculty toward research and publication ethics (*70*, *71*). However, we believe that an alternative approach might be one that focuses on educating the graduate student mentors, who represent the next generation of faculty, as education and training on authorship meets their immediate and future needs. Increasingly, a collection of peer-reviewed publications is being viewed as the expected outcome of doctoral education. Indeed, there has even been a recent effort to quantify this expected output in the natural and biomedical sciences (*72*). Graduate students are frequently tasked with the responsibility of mentoring undergraduate research students and often appear to be involved in authorship decision-making. As a result, mentoring programs for graduate students that include training on RCR issues have the potential to widely impact and positively transform the culture of scientific research. This "stealth" approach, focused on educating graduate students in the critical issues surrounding the assignment of credit, has the potential to impact the next generation of faculty, giving them the knowledge and self-confidence to engage in open and honest conversations with their protégés.

### Group Meetings and Journal Clubs

Many research groups hold regular group meetings and journal clubs (*73*). Interested faculty could assign and discuss select cases with their research teams in group meetings ensuring that all team members are on the same page when it comes to authorship criteria and order. Several excellent textbooks (*74–76*), websites (*77*, *78*), and even some journal articles include real and hypothetical case studies focused on authorship credit, responsibility, and order suitable for use with graduate and undergraduate student audiences (*79–82*). One of the unique characteristics of this ACS Symposium series volume is that it contains a series of case studies focused explicitly on some of the challenging issues related to the assignment of credit, authorship, and inventorship. We hope these case studies will be used in fora like group meetings for education and training on authorship.

### Written Contracts

No one likes conflict, but the conflict in authorship decision-making could be minimized if not eliminated if the group leader discussed the criteria for authorship at the start of the research project with all team members and if the team outlined his or her agreement in writing. Moreover, if throughout the process of publication, when changes to the byline need to be made, authorship teams would be willing to revisit and revise their written agreements, and the negotiation of authorship would be more objective and less stressful for all participants.

The use of negotiated written contracts is not a new idea in education (*83*). Indeed, research learning contracts (RLCs) are frequently used in collaborations between faculty and undergraduates to outline expectations and needs and to maximize the likelihood of productive collaborative experiences (*84–88*). Moreover, individual development plans (IDPs) are increasingly used in graduate and postdoctoral training in STEM research. Their use is strongly encouraged for students working on NIH supported research (*89*). Both RLCs and IDPs, when appropriately used, are intended to be revisited, as needed, by the mentor and protégé, and to be signed signaling agreement and commitment to the terms of the documents. The inclusion of criteria for authorship would make the decision-making process transparent, uniform, and objective.

## Closing Words

While we humbly recognize that this work has focused on a single research group in one STEM discipline at one research university, we challenge readers to consider that the diversity of experiences and opinions captured in this study may reflect the experience of other research teams, perhaps even teams of which you have been a member or teams that you lead. Faculty need to be prepared to take the lead and actively and openly discuss research ethics issues including authorship with their junior colleagues. Based on our work, faculty are waiting for their students to ask, and students are waiting for their faculty to tell them. So, everyone is waiting, and no one is leading. Faculty need to remember that there's an inherent power differential in their relationship with students, including postdoctoral, graduate, and undergraduate, which makes it challenging for student scientists to speak up, even when they know that they should. To avoid confusion and misunderstanding and to ensure that everyone is on the same page, we would encourage faculty to consider engaging their whole laboratory as a team in open authorship decision-making discussions in which they explicitly disclose their criteria and their decision-making process.

## Acknowledgments

The authors wish to thank the Office of the Provost at Northeastern University for funding this research project. We wish to acknowledge Aneri Pattani who helped design and refine the interview scripts used in this work and set up the original codebook used in this study. We also wish to thank all of the graduate and undergraduate research students who participated in our study for their candor, enthusiasm, and support. Lastly, we want to thank Rein Kirss, Vaso Lykourinou, Gautam Bhattacharyya, and Lauren Abbott for helpful comments and suggestions on various drafts of this manuscript.

# Appendix A. Undergraduate Research Student Script

### Introduction

I would like to learn about your experience of the science ethics training in your undergraduate research experience and more specifically more specifically your experiences related to authorship.

### Engagement Questions

1. What are you working on and what have you learned?
2. When do you anticipate being graduated?
3. What is your academic major?
4. What is your current year of study?
5. What are your current career goals?
6. What are you planning to do after graduation?
7. I would like to learn a bit about your undergraduate research experience.

    a. What motivated you to get involved in undergraduate research?
    b. What did you hope to get out of your undergraduate research?
    c. What motivated you to work in your current research experience?
    d. Why did you choose your current advisor?

8. Did you have any specific interest in authorship opportunities?
9. Have you done undergraduate research before? Tell me a bit about your work.
10. Have you ever been involved in a project that resulted in a publication?

    a. Did it have your name on it? If not, why was your name not included?

### Exploration Questions

1. Did you ever receive any formal research ethics training? Was this a course? Workshop?

    <u>If the student did not receive any formal training:</u>

    a. Did you ever receive any informal research ethics training in which issues related to authorship, confidentiality, openness, conflict of interest, fabrication/falsification of data, etc. were discussed?
    b. What information was discussed with you and what was the context of the discussion? I would like to know with whom you spoke and when and why the topic(s) was/were discussed.

a. <u>If the student received any training formal or informal:</u>

    a. What information was covered in your training?
    b. Did you find the training helpful and how? Did it fall short? In what ways??

2. Do you expect to be an author or obtain a patent for any work you have done as an undergraduate research student in your current position?
3. What are the requirements for authorship in your research laboratory? <u>If the student knows the requirements for authorship in their lab:</u>

    a. When and how did you learn about the requirements for authorship in your research laboratory?

a. <u>If the student doesn't know the requirements for authorship in their lab:</u>

    a. Are there reasons or obstacles that have prevented you from learning about the requirements for authorship in your laboratory?
    b. What are some things we could do to help remove these obstacles and help you obtain answers to your questions about authorship?

4. Do you think there should be set standards for authorship or that it should be more dependent on each situation? If standards, what should those standards be?
5. Imagine that you are a member of a team designing science ethics training about authorship for undergraduate research students.

    a. What are the factors that you believe the committee should consider in designing this training? These factors could include the mechanism for the training (face-to-face/online/single-shot-workshop, case-study based or anything else you can think of.
    b. What are the things that you believe would help students learn the key science ethics issues?

**Exit Questions**

1. Is this interview what you expected?
2. Did you learn anything new about science ethics and science ethics training by being participating today? What did you learn?
3. Is there anything else we haven't discussed yet that you think is important for me to know about as we consider designing and implementing science ethics training programs for undergraduates?

# Appendix B. Graduate Student Script

### Introduction

I would like to learn about your experience in science ethics training and your relationships with both undergraduate research students and your advisor. More specifically your experiences related to authorship.

### Engagement Questions

1. What are you working on in your current lab?
2. When do you anticipate being graduated from graduate school?
3. What is your academic discipline?
4. When did you begin your PhD?
5. What are you planning to do after graduation?
6. How many undergraduates worked with you?
7. Did you participate in undergraduate research?
8. What has been your experience with authorship to date?
9. Have you ever been involved in a project that resulted in a publication?

    a. Did it have your name on it? If not, why was your name not included?

### Exploration Questions

2. Have you received any formal research ethics training? Was this a course? Workshop?

    If the graduate student did not receive any formal training:

    a. Did you ever receive any informal research ethics training in which issues related to authorship, confidentiality, openness, conflict of interest, fabrication/falsification of data, etc. were discussed?
    b. What information was discussed with you and what was the context of the discussion? I would like to know with whom you spoke and when and why the topic(s) was/were discussed.

a. If the graduate student received any training formal or informal:

    a. What information was covered in your training?
    b. Did you find the training helpful and how? Did it fall short? In what ways?

3 Do you expect to be an author or obtain a patent for any work you have done as a graduate research student in your current position?

4 What are the requirements for authorship in your research laboratory?

If the graduate student knows the requirements for authorship in their lab:

a. When and how did you learn about the requirements for authorship in your research laboratory?

a. If the graduate student doesn't know the requirements for authorship in their lab:

a. Are there reasons or obstacles that have prevented you from learning about the requirements for authorship in your laboratory?
b. What are some things we could do to help remove these obstacles and help you obtain answers to your questions about authorship?
c. Did your advisor explain the rules of authorship for the lab?

5 Do you think there should be set standards for authorship or that it should be more dependent on each situation? If standards, what should those standards be?
6 Do you have any role in decision-making about authorship with regard to the undergraduates you work with?

If the graduate student does:

a. What is your role/responsibility?
b. What are the factors that you consider in deciding on student authorship?

a. If the graduate student does not:

a. Are you aware of how your advisor decides on student authorship?
b. Has your advisor asked you about the undergraduate student's work when making authorship decisions?
c. What questions has your advisor asked about the undergraduate student's work?

7 Imagine that you are a member of a team designing science ethics training about authorship for undergraduate research students.

a. What are the factors that you believe the committee should consider in designing this training? These factors could include the mechanism for the training (face-to-face/online/single-shot-workshop, case-study based or anything else you can think of.
b. What are the things that you believe would help undergraduate students learn the key science ethics issues?

**Exit Questions**

1. Is this interview what you expected?
2. Did you learn anything new about science ethics and science ethics training by being participating today? What did you learn?
3. Is there anything else we haven't discussed yet that you think is important for me to know about as we consider designing and implementing science ethics training programs for undergraduates and graduates?

## References


1. Hirsch, J. E. An Index to Quantify an Individual's Scientific Research Output. *Proc. Natl. Acad. Sci. U.S.A.* **2005**, *102*, 16569–16572.
2. Osborne, J. W.; Holland, A. What is Authorship, and What Should It Be? A Survey of Prominent Guidelines for Determining Authorship in Scientific Publications. *PARE* **2009**, *14*, 19.
3. Solomon, J. Programmers, Professors, and Parasites: Credit and Co-Authorship in Computer Science. *Sci. Eng. Ethics* **2009**, *15*, 467–489.
4. Marusic, A.; Bosnjak, L.; Jeroncic, A. A Systematic Review of Research on the Meaning, Ethics and Practices of Authorship across Scholarly Disciplines. *PLoS One* **2011**, *6*, e23477.
5. Jennings, M. M.; El-adaway, I. H. Ethical Issues in Multiple-Authored and Mentor-Supervised Publications. *J. Prof. Issues Eng. Educ. Pract.* **2012**, *138*, 37–47.
6. Seeman, J. I.; House, M. C. Influences on Authorship Issues: An Evaluation of Receiving, Not Receiving, and Rejecting Credit. *Account. Res.* **2010**, *17*, 176–197.
7. Royal Society of Chemistry Author Responsibilities. *Ethical Guidelines and Code of Conduct for Authors*. http://www.rsc.org/journals-books-databases/journal-authors-reviewers/author-responsibilities/ (accessed May 10, 2018).
8. *Editors of the Publications Division of the American Chemical Society Ethical Guidelines to Publication of Chemical Research*. http://pubs.acs.org/userimages/ContentEditor/1218054468605/ethics.pdf (accessed May 10, 2018).
9. *Committee on Publication Ethics COPE*. https://publicationethics.org/ (accessed May 10, 2018).
10. *International Committee of Medical Journal Editors ICMJE International Committee of Medical Journal Editors*. http://www.icmje.org/ (accessed May 10, 2018).
11. Clement, T. P. Authorship Matrix: A Rational Approach to Quantify Individual Contributions and Responsibilities in Multi-Author Scientific Articles. *Sci. Eng. Ethics* **2014**, *20*, 345–361.
12. Clement, T. P. Who Are Coauthors and What Should Be Their Responsibilities? *Environ. Sci. Technol.* **2015**, *49*, 3265–3266.
13. McNutt, M.; Bradford, M.; Drazen, J.; Hanson, R. B.; Howard, B.; Jamieson, K. H.; Kiermer, V.; Magoulias, M.; Marcus, E.; Pope, B. K.;

Schekman, R.; Swaminathan, S.; Stang, P.; Verma, I. Transparency In Authors' Contributions And Responsibilities To Promote Integrity In Scientific Publication. *bioRxiv* **2017**.
14. Smith, E.; Williams-Jones, B. Authorship and Responsibility in Health Sciences Research: A Review of Procedures for Fairly Allocating Authorship in Multi-Author Studies. *Sci. Eng. Ethics* **2012**, *18*, 199–212.
15. Verdan, A. M.; Ingallinera, J. T.; Bhattacharyya, G. Scientific Norms and Ethical Misconduct: Research Towards the Design of a Course in Scientific Ethics. *Chem. Educ. Res. Pract.* **2010**, *11*, 118–123.
16. Swazey, J. P.; Anderson, M. S.; Lewis, K. S.; Louis, K. S. Ethical Problems in Academic Research. *Am. Sci.* **1993**, *81*, 542–553.
17. Rose, M.; Fischer, K. Policies and Perspectives on Authorship. *Sci. Eng. Ethics* **1995**, *1*, 361–370.
18. *APA Science Student Council A Graduate Student's Guide to Determining Authorship Credit and Authorship Order.* http://www.apa.org/science/leadership/students/authorship-paper.pdf (accessed May 10, 2018).
19. Eastwood, S.; Derish, P.; Leash, E.; Ordway, S. Ethical Issues in Biomedical Research: Perceptions and Practices of Postdoctoral Research Fellows Responding to a Survey. *Sci. Eng. Ethics* **1996**, *2*, 89–114.
20. Tarnow, E. The authorship list in science: Junior physicists' perceptions of who appears and why. *Sci. Eng. Ethics* **1999**, *5*, 73–88.
21. Mabrouk, P. A.; Peters, K. Student Perspectives on Undergraduate Research Experiences in Chemistry and Biology. *CUR Q.* **2000** (September), 25–33.
22. Heitman, E.; Olsen, C. H.; Anestidou, L.; Bulger, R. E. New Graduate Students' Baseline Knowledge of the Responsible Conduct of Research. *Acad. Med.* **2007**, *82*, 838–845.
23. Yang, S. C. Ethical Academic Judgments and Behaviors: Applying a Multidimensional Ethics Scale to Measure the Ethical Academic Behavior of Graduate Students. *Ethics Behav.* **2012**, *22*, 281–296.
24. Arda, B. Publication Ethics from the Perspective of Ph.D. Students of Health Sciences: A Limited Experience. *Sci. Eng. Ethics* **2012**, *18*, 213–222.
25. Brown, S.; Kalichman, M. W. Effects of Training in the Responsible Conduct of Research: A Survey of Graduate Students in Experimental Sciences. *Sci. Eng. Ethics* **1998**, *4*, 487–498.
26. Plemmons, D. K.; Body, S. A.; Kalichman, M. W. Student Perceptions of the Effectiveness of Education in the Responsible Conduct of Research. *Sci. Eng. Ethics* **2006**, *12*, 571–582.
27. Powell, S. T.; Allison, M. A.; Kalichman, M. W. Effectiveness of a Responsible Conduct of Research Course: A Preliminary Study. *Sci. Eng. Ethics* **2007**, *13*, 249–264.
28. Antes, A. L.; Murphy, S. T.; Waples, E. P.; Mumford, M. D.; Brown, R. P.; Connelly, S.; Devenport, L. D. A Meta-Analysis of Ethics Instruction Effectiveness in the Sciences. *Ethics Behav.* **2009**, *19*, 379–402.
29. Plemmons, D. K.; Kalichman, M. W. Reported Goals of Instructors of Responsible Conduct of Research for Teaching of Skills. *J. Empir. Res. Hum. Res. Ethics* **2013**, *8*, 95–103.

30. May, D. R.; Luth, M. T. The Effectiveness of Ethics Education: A Quasi-Experimental Field Study. *Sci. Eng. Ethics* **2013**, *19*, 545–568.
31. Mabrouk, P. A. An Investigation of the Evolution of High School and Undergraduate Student Researchers' Understanding of Key Science Ethics Concepts. *J. Coll. Sci. Teach.* **2013**, *43*, 91–99.
32. Mabrouk, P. A. What Knowledge of Responsible Conduct of Research Do Undergraduates Bring to Their Undergraduate Research Experiences? *J. Chem. Educ.* **2016**, *93*, 46–55.
33. Anderson, D. D.; Shore, W. J. Ethical Issues and Concerns Associated With Mentoring Undergraduate Students. *Ethics Behav.* **2008**, *18*, 1–25.
34. Burks, R. L.; Chumchal, M. M. To Co-Author or Not to Co-Author: How to Write, Publish, and Negotiate Issues of Authorship with Undergraduate Research Students. *Science Signaling* **2009**, *2*, tr3.
35. Welfare, L. E.; Sackett, C. R. The Authorship Determination Process in Student–Faculty Collaborative Research. *J. Couns. Dev.* **2011**, *89*, 479–487.
36. Maurer, T. Guidelines for Authorship Credit, Order, and Co-Inquirer Learning in Collaborative Faculty-Student SoTL Projects. *Teaching Learning Inquiry* **2017**, *5*, 1–17.
37. Fine, M. A.; Kurdek, L. A. Reflections on Determining Authorship Credit and Authorship Order on Faculty-Student Collaborations. *Am. Psychol.* **1993**, *48*, 1141–1147.
38. Gibbs, G. *Analyzing Qualitative Data*; Sage: Los Angeles, 2007; Vol. 6, p 160.
39. Wertz, F. J.; Charmaz, K.; McMullen, L. M.; Josselson, R.; Anderson, R.; McSpadden, E. *Five Ways of Doing Qualitative Analysis*; The Guilford Press: New York, 2011.
40. Saldana, J. *The Coding Manual for Qualitative Researchers*, 2nd ed.; Sage: Los Angeles, 2013.
41. Corbin, J.; Strauss, A., *Basics of Qualitative Research: Techniques and Procedures for Developing Grounded Theory*, 4th ed.; SAGE Publications, Inc.: Thousand Oaks, 2015; p 456.
42. Abbott, L.; Andes, A.; Pattani, A.; Mabrouk, P. A. A Theoretical Model for Authorship Decision-Making in Undergraduate Research Partnerships at Research Universities. *Sci. Eng. Ethics* **2018**in preparation.
43. Wenger, E., *Communities of Practice.* Cambridge University Press: Cambridge, UK, 1998; p 318.
44. Allen, T. D.; Poteet, M. L.; Burroughs, S. M. The Mentor's Perspective: A Qualitative Inquiry and Future Research Agenda. *J. Vocat. Behav.* **1997**, *51*, 70–89.
45. Dolan, E.; Johnson, D. Toward a Holistic View of Undergraduate Research Experiences: An Exploratory Study of Impact on Graduate/Postdoctoral Mentors. *J. Sci. Educ. Technol.* **2009**, *18*, 487.
46. Verdan, A. M. *Finding a New Continent Versus Mapping All the Rivers: Recognition, Ownership, and the Scientific Epistemological Development of Practicing Scientists and Engineers Dissertation.* Clemson University, Clemson, South Carolina, 2012.

47. Merkel, C. A.; Baker, S. M. *How to Mentor Undergraduate Researchers*; the Council on Undergraduate Research: Washington, DC, 2002; p 29.
48. Ealy, J. B.; Kvarta, V. Mentoring an Undergraduate Research Student in the Structural and Nonstructural Properties of Drugs. *J. Chem. Educ.* **2006**, *83*, 1779–1783.
49. Pfund, C.; Pribbenow, C. M.; Branchaw, J.; Lauffer, S. M.; Handelsman, J. The Merits of Training Mentors. *Science (Washington, D. C., 1883-)* **2006**, *311*, 473–474.
50. Dolan, E. L.; Johnson, D. The Undergraduate–Postgraduate–Faculty Triad: Unique Functions and Tensions Associated with Undergraduate Research Experiences at Research Universities. *CBE Life Sci. Educ.* **2010**, *9*, 543–553.
51. Revelo, R. A.; Loui, M. C. A Developmental Model of Research Mentoring. *College Teaching* **2016**, *64*, 119–129.
52. Mabrouk, P. A. Development and Implementation of an Effective Graduate Student Mentoring Program for in Support of Undergraduate Research Experiences. *PURM* **2017**, *6.1*, 1–22.
53. Mabrouk, P. A. Research Skills & Ethics - A Graduate Course Empowering Graduate Students for Productive Research Careers in Graduate School and Beyond. *J. Chem. Educ.* **2001**, *78*, 1628–1631.
54. Allen, T. D.; Eby, L. T. Factors Related to Mentor Reports of Mentoring Functions Provided: Gender and Relational Characteristics. *Sex Roles* **2004**, *50*, 129–139.
55. Koberg, C. S.; Boss, R. W.; Goodman, E. Factors and Outcomes Associated with Mentoring among Health-Care Professionals. *J. Vocat. Behav.* **1998**, *53*, 58–72.
56. Lockwood, P. "Someone Like Me Can Be Successful": Do College Students Need Same Gender Role Models? *Psychol. Women Q.* **2006**, *30*, 36–46.
57. Ragins, B. R.; McFarlin, D. B. Perceptions of Mentor Roles in Cross-Gender Mentoring Relationships. *J. Vocat. Behav.* **1990**, *37*, 321–339.
58. Thomas, K. M.; Hu, C.; Gewin, A. G.; Bingham, K.; Yanchus, N. The Roles of Protégé Race, Gender, and Proactive Socialization Attempts on Peer Mentoring. *Adv. Develop. Human. Resour.* **2005**, *7*, 540–555.
59. Thomas, D. A. The Impact of Race on Managers' Experiences of Developmental Relationships (Mentoring and Sponsorship): An Intra-Organizational Study. *J. Organ. Behav.* **1990**, *11*, 479–492.
60. *Lockheed Martin Corporation Voicing Our Values 2017.* https://lockheedmartin.com/us/who-we-are/ethics/2017_Awareness/index.html?&lang=en (accessed May 10, 2018).
61. Bowen, G. M.; Roth, W.-M. The "Socialization" and Enculturation of Ecologists in Formal and Informal Settings. *Electron. J. Sci. Educ.* **2002**, *6*, 1–26.
62. Haynes, A. B.; Weiser, T. G.; Berry, W. R.; Lipsitz, S. R.; Breizat, A.-H. S.; Dellinger, E. P.; Herbosa, T.; Joseph, S.; Kibatala, P. L.; Lapitan, M. C. M.; Merry, A. F.; Moorthy, K.; Reznick, R. K.; Taylor, B.; Gawande, A. A. A Surgical Safety Checklist to Reduce Morbidity and Mortality in a Global Population. *New Engl. J. Med.* **2009**, *360*, 491–499.

63. Spector, J. M.; Agrawal, P.; Kodkany, B.; Lipsitz, S.; Lashoher, A.; Dziekan, G.; Bahl, R.; Merialdi, M.; Mathai, M.; Lemer, C.; Gawande, A. Improving Quality of Care for Maternal and Newborn Health: Prospective Pilot Study of the WHO Safe Childbirth Checklist Program. *Plos One* **2012**, *7*, e35151.
64. Perry, W.; Bagheri Nejad, S.; Tuomisto, K.; Kara, N.; Roos, N.; Dilip, T.; Hirschhorn, L.; Larizgoitia, I.; Semrau, K.; Mathai, M.; Dhingra-Kumar, N. Implementing the WHO Safe Childbirth Checklist: Lessons From a Global Collaboration. *BMJ Global Health* **2017**, *2*, e000241.
65. Spector, J.; Lashoher, A.; Agrawal, P.; Lemer, C.; Dziekan, G.; Bahl, R.; Mathai, M.; Merialdi, M.; Berry, W.; A Gawande, A. Designing the WHO Safe Childbirth Checklist Program to Improve Quality of Care at Childbirth. *Int. J. Gynecol. Obstet.* **2013**, *122*, 164–168.
66. *World Health Organization WHO Safe Childbirth Checklist.* http://www.who.int/patientsafety/implementation/checklists/childbirth/en/ (accessed May 10, 2018).
67. *World Health Organization WHO Surgical Safety Checklist.* http://www.who.int/patientsafety/topics/safe-surgery/checklist/en/ (accessed May 10, 2018).
68. Clay-Williams, R.; Colligan, L. Back to Basics: Checklists in Aviation and Healthcare. *BMJ Quality Safety* **2015**, *24*, 428–431.
69. Fourcade, A.; Blache, J.-L.; Grenier, C.; Bourgain, J.-L.; Minvielle, E. Barriers to staff adoption of a surgical safety checklist. *BMJ Quality Safety* **2012**, *21*, 191–197.
70. Kim, S. Y. Students' Evaluation of a Team-based Course on Research and Publication Ethics: Attitude Change in Medical School Graduate Students. *J. Educ. Eval. Health Prof.* **2008**, *5*, 3.
71. Ju, Y.-S. Evaluation of a Team-Based Learning Tutor Training Workshop on Research and Publication Ethics by Faculty and Staff Participants. *J. Educ. Eval. Health Prof.* **2009**, *6*, 5.
72. Hagen, N. T. Deconstructing Doctoral Dissertations: How Many Papers Does it Take to Make a Ph.D.? *Scientometrics* **2010**, *85*, 567–579.
73. Deenadayalan, Y.; Grimmer-Somers, K.; Prior, M.; Kumar, S. How to Run an Effective Journal Club: A Systematic Review. *J. Eval. Clin. Pract.* **2008**, *14*, 898–911.
74. Kovac, J. *The Ethical Chemist*; Prentice Hall: Upper Saddle River, NJ, 2003; p 122.
75. Macrina, F. Teaching Authorship and Publication Practices in the Biomedical and Life Sciences. *Sci. Eng. Ethics* **2011**, *17*, 341–354.
76. D'Angelo, J. *Ethics in Science: Ethical Misconduct in Scientific Research*; CRC Press: Boca Raton, FL, 2012; p 112.
77. *National Academy of Sciences The Online Ethics Center for Engineering and Science.* http://www.onlineethics.org/ (accessed May 10, 2018).
78. *Committee on Publication Ethics Cases.* https://publicationethics.org/cases (accessed May 10, 2018).
79. Shachter, A. M. Integrating Ethics in Science into a Summer Undergraduate Research Program. *J. Chem. Educ.* **2003**, *80*, 507–512.

80. Niece, B. K. Who Is Responsible for a Fraud: An Exercise Examining Research Misconduct and the Obligations of Authorship through Case Studies. *J. Chem. Educ.* **2005**, *82*, 1521.
81. Fisher, E. R.; Levinger, N. E. A Directed Framework for Integrating Ethics into Chemistry Curricula and Programs Using Real and Fictional Case Studies. *J. Chem. Educ.* **2008**, *85*, 796–801.
82. Montes, I.; Padilla, A.; Maldonado, A.; Negretti, S. Student-Centered Use of Case Studies Incorporating Oral and Writing Skills To Explore Scientific Ethical Misconduct. *J. Chem. Educ.* **2009**, *86*, 936–939.
83. Albert, T.; Wager, E. *How to Handle Authorship Disputes: A Guide for New Researchers*; Committee on Publication Ethics (internet), 2003; pp 32−34.
84. Knowles, M. S. *Self-Directed Learning*; Cambridge: New York, 1975; p 400.
85. Knowles, M. S. *Using Learning Contracts*; Jossey-Bass, Inc.: San Francisco, 1986; p 262.
86. Mabrouk, P. A. Research Learning Contracts: A Useful Tool for Facilitating Successful UR Experiences. *CUR Q.* **2003**, *24*, 26–30.
87. Rye, K. J.-B. Perceived Benefits of the Use of Learning Contracts to Guide Clinical Education in Respiratory Care Students. *Respir. Care* **2008**, *53*, 1475–1481.
88. Bone, Z. Using a Learning Contract to Introduce Undergraduates to Research Projects. *EJBRM* **2014**, *12*, 115–123.
89. Office of Policy for Extramural Research Administration. Division of Grants Policy *Revised Policy: Descriptions on the Use of Individual Development Plans (IDPs) for Graduate Students and Postdoctoral Researchers Required in Annual Progress Reports beginning October 1, 2014*; National Institutes of Health.

## Chapter 14

# Case Study: Authorship with Students in the Research Group at the Research University

**John D'Angelo***

**Division of Chemistry, Alfred University, Alfred, New York14802, United States**
***E-mail: dangelo@alfred.edu.**

Authorship is sometimes complicated, especially when colleagues disagree about when to publish. This can particularly be an issue between students and their mentors. Care must be taken to ensure that worthy coauthors, especially undergraduates, are not snubbed. Herein are cases that explore the exclusion of potentially or questionably worthy authors and disagreements about when something should be published.

## Imagine This…

Frank Hawk is about to graduate and, as he finishes his Ph.D. work, he trains a new graduate student, Callie Sampson, to take the project to the next stages. A few years go by, and Frank has settled into his independent career as a faculty member. He sees Callie at a conference, and the two get to chatting. At one point, he asks Callie, "What happened to our project, I'm surprised you and Jones (their mentor) haven't published yet."

Callie responds in a frustrated tone, "Oh, don't get me started. Every time I think I'm done, the boss gives me five more experiments to try, saying, 'It'll make the paper bigger, and bigger is better.'"

Frank contemplates this for a moment and says to her, "Hey, you know what? Since I technically started the project and handed it off to you, maybe we should just write it together. We'll add Jones as an author, of course, but, we should be able to publish **our** work, right?"

Callie considers Frank's offer for a moment and says, "Frank, that'd be great. I really want this postdoc position at Harvard, but I'll never get it without at least one more paper."

*Some points to ponder:*

- Are Frank and Callie right? Is this *their* project and work to publish?
- When, if at all, should Frank and Callie let Jones know of their plan?
- If you were Jones, what would you do upon receiving such a draft?

Issues where the opinion of a student or other research associate differs from that of the PI on the publishability of the work are not uncommon. In one case, Mavi Camarasa, who once worked in the laboratories of Daniel Bachiller, felt the work was more publishable than Bachiller. Their lab was apparently the first to correct the most common cystic fibrosis mutation in stem cells derived from a patient, and this sort of discovery would potentially be of benefit to afflicted persons or carriers of this terrible disease. Camarasa claimed that, much to her chagrin, Bachiller insisted that the project was not yet complete and that the publication must wait. A a few years passed, Camarasa left the lab, and a large portion of this work was scooped (meaning, another group published a report of their own, similar work). Camarasa hatched a plan to write the paper, get it accepted, and then tell Bachiller, despite the fact that she was no longer a member of his lab. Her first attempt at publication was rejected. After switching target journals, she finally told Bachiller of her plan. He responded, not by telling her to stop altogether but by insisting that she submit a more complete paper to a higher-profile journal than her revised target. Camarasa rejected this idea and went ahead with her original plan; the paper was accepted and subsequently published. Bachiller found out and engaged in a dispute over the work.

*Some additional points to ponder:*

- Should Bachiller have ordered one of the current workers to do the additional work necessary to expand the paper as he desires?
- Was Bachiller unreasonable with his demands?
- Should Camarasa have simply written the more complete paper as Bachiller suggested, even though Camarasa had been removed from the project for a few years by then?
- What should the journal do with this paper?
- Did Camarasa have any right to do as she did?
- Does it matter that Camarasa is no longer part of the lab?

To find out the fate of this paper, see the source below:

Source:

Han, A.P. Retraction Watch. *Author of retracted gene editing paper alleges "bullying" by former PI.* June 12, 2017. http://retractionwatch.com/2017/06/12/author-retracted-gene-editing-paper-alleges-bullying-former-pi/#more-50485 (accessed January 22, 2018).

## Now, Imagine This Scenario…

Felicity Brick just got back from her REU program (Research Experience for Undergraduates, a National Science Foundation-supported summer research program), and she is talking with her friend, Jess Sinclair, who had done summer researchwith one of their professors at their home institution, a PUI (Primarily Undergraduate Institution). They are exchanging stories about summer experiences when Felicity receives a text message from the graduate student with whom she worked most closely, and, by a strange coincidence, Jess gets one at the same time from the professor with whom she worked. After reading their texts, they exchange excited news, Felicity saying, "They're getting published" while Jess simultaneousy blurts out, "I'm getting published". Jess then says, almost sympathetically, "Oh, Felicity, you did way more work on your project than I did on mine. Why aren't you on the author list too?"

"Oh, I don't mind; the group is huge and like half the grad students are on the paper," Felicity responds, continuing, "Plus, I was just a summer student…I bet someone repeated my results dozens of times, even since I left. They probably didn't even use my actual data, even if they were using the optimizations I made to the protocol using that trick we learned in Biochem with Professor Hendey to enhance signal to noise. The data we started getting was so much better than anything before it. They even cited the lab manual Hendy wrote for our class."

Jess frowns and says "I don't think it matters that the group is big and that they repeated your results. It's not ethical to exclude you if you contributed to the work, and it really sounds to me like you did."

*Some points to ponder*

- Is Felicity right to be so easy-going, or is Jess right that Felicity should be included?
- Should Hendey have been offered coauthorhip?

Such contributions of undergraduate researchers are often "at risk" of going unacknowledged. A few years ago, an individual claiming to be an undergraduate researcher posted an unsubstantiated comment on reddit that they were denied authorship on a paper, despite the fact that they made, from their perspective, major contributions to the work and wrote at least the introduction to the paper. Although there is no reason to believe the case is, in fact, real, in the absence of evidence of it being fabricated, let us assume its veracity since it is illustrative of several important issues.

According to this student, their adviser for the research project (who was a postdoc and **not** the PI), said that authorship would not be given because "…despite the significant work you are putting into it, given the already large number of coauthors…" (see source below) The student was also given a promise of being listed as a coauthor *next time*. According to the post, graduate students in the lab were supportive of the student, as were most of the responders to the post on reddit. In particular, readers pointed out that the number of authors should not impact this decision at all. Also, readers pointed out that this decision should not be the postdoc's to make; rather, the PI should make such a decision. Both reader points are valid. Taking the comments from not only the graduate students but also the postdoc at face value, the undergraduate student was not inflating their role. The postdoc's comments, in particular, were very suggestive that the student made a significant contribution.

It is noteworthy that one of the things compelling the student to not fight this and bring it to the attention of the PI is the anticipation or plan to continue working in this lab for the next few years. The student is worried that this sort of fight would spoil a working relationship and potentially cost them a letter of recommendation. This fear, though reasonable is grossly unfair. Much is (rightfully) made of the virtual stranglehold an advisor has over graduate students and how graduate students are the academic equivalent of an indentured servant. Undergraduates have it at least as hard.

*Some additional points to ponder:*

- What else could the student have done?
- If the graduate students really thought this student deserved authorship, are they ethically bound to speak up?
- Is the PI at fault for not being more involved?

*Note: A postdoc is someone who has already completed Ph.D. studies and is employed to work in the lab, typically as a researcher, although some post doc appointments include teaching as well. Such appointments typically last 1-3 years. Postdocs are often leaders in the labs in which they work and are given more responsibilities than a typical graduate student but do not experience the same demands as a faculty person (i.e. they are not PIs).*

## Source

Mechanical_Toaster. Reddit.com: Ask Academia. *I've written most of a paper and was denied co-authorship;* 2015. https://www.reddit.com/r/AskAcademia/comments/2elcqk/ive_written_most_of_a_paper_and_was_denied/ (accessed April 26, 2018).

## Chapter 15

# Patterns in Authorship: Lessons in Diversity and Justice

**Arthur Greenberg***

**Department of Chemistry, University of New Hampshire, Durham, New Hampshire 03824, United States**
***E-mail: art.greenberg@unh.edu.**

A very brief outline of some of the significant changes in journal publication and authorship practices over the last half century is presented. Among these are the sheer increase in scientists, the world-wide expansion of sources of papers, the movement toward multi-disciplinary research projects and authorships, and the increase in the number of journals and symposia, a very few of which are of inferior quality. This is followed by a brief discussion of modern standards for inclusion in the authors' byline or acknowledgements and a discussion of the order of authors in the byline. The concluding section examines the increasing diversity in the gender, race, and ethnicity of contributing authors and acknowledges the progress that remains to be made.

## Introduction

Careers and friendships are made and broken in collaborative authorships. Questions of who should or should not be included in the author byline of a research paper can sometimes be quite subtle. Should a person's name appear among the co-authors or in the acknowledgements section? What should the order of authors be, especially the prized positions of first author and last author? Science has become much more interdisciplinary in recent times. Universities increasingly name "centers", "institutes", and "schools" in order to recognize this fact and to garner attention. Increasingly, universities are becoming involved in "big science." Whereas, decades ago, each author on a paper was expected to be an expert concerning virtually every point in the paper, this is less of an expectation in a large interdisciplinary group. As winning research grants and obtaining tenure and promotion at colleges and universities become ever more competitive, defining authorship and attributing it logically and equitably becomes even more critical. Our profession has become more diverse in gender, race, and ethnicity. How can we best encourage the development of a diverse and talented scientific culture? The present chapter lightly touches on modern criteria for authorship, movements for clarification of the contributions of contributing authors, the underrepresentation of women and racial and ethnic groups in key author positions, and suggestions for improvement.

I began my graduate study in chemistry in 1967, over half a century ago. The shifts in the landscape of publications have been tectonic during this period. In my entering graduate class of roughly 35 there were 2 women, apparently the first two admitted to this particular graduate program. My recollection was of one Hispanic student and no African-American students. Post-doctoral researchers in the department were mostly European, Japanese, and Indian. My research area could be classified as physical organic chemistry, very much a recognized, well-defined discipline at that time. Maintaining some degree of currency with the chemical literature was a demanding, but seemingly finite task. *Journal of the American Chemical Society*, *Journal of Organic Chemistry*, *Journal of the Chemical Society (London)*, *Angewandte Chemie International Edition in English*, *Tetrahedron*, *Tetrahedron Letters*, and *Chemical Communications* were among the journals that provided excellent currency for a native speaker of English, although passing chemistry language exams in French and German extended access to additional high-quality journals. Although an exam in chemical Russian was also available at Princeton, there were very few "takers". These journals also provided a degree of coverage of potential questions for the feared monthly cumulative exams. Although my thesis included a few references to contributions from Japanese and Indian authors, there were none to authors based in China and "third-world" countries. A quick glance at Table 1 demonstrates the increasingly world-wide distribution of sources of chemical literature between 1939 (when Chemical Abstracting Service started listing these data) and 2004 (CAS terminated this survey in 2007). This table has relevance to the present chapter since, not only do authorship customs vary with fields, but they also vary in countries having different academic traditions and customs. This interesting table has its own historical and cultural artifacts (imagine listing Canada, India, and Australia along with the United Kingdom as the British Empire).

In 2009, the National Academy Press published the third edition of its superb pamphlet, *On Being a Scientist. A Guide to Responsible Conduct in Research* (*1*). In addition to providing guidelines for best practices, case studies (historical and hypothetical) are provided for group discussion. One section of the pamphlet is titled "Authorship and the Allocation of Credit" (*1*).

**Table 1. Sources (Percent) of Literature Abstracted in *Chemical Abstracts*[a,b]**

| *Nation* | *1939* | *1956* | *1972* | *1993* | *2004* |
|---|---|---|---|---|---|
| USA | 27.7% | 28.4% | 28.0% | 28.1% | 23.2% |
| Japan | 4.4 | 10.4 | 7.9 | 13.3 | 11.8 |
| China | | | | 4.3[c] | 12.0[c] |
| Germany | 18.7 | 8.4 | | 7.5[d] | 6.2[e] |
| France | 9.1 | 6.0 | 4.4 | 4.7 | 3.7 |
| Italy | 3.0 | 4.1 | 1.9 | 2.6 | 3.1 |
| UK | 14.1[f] | 13.6[f] | 6.4 | 5.7 | 4.6 |
| Canada | *f* | *f* | 2.8 | 3.2 | 2.5 |
| India | *f* | *f* | 2.5 | 2.6 | 2.6 |
| Australia | *f* | *f* | 1.2 | 1.4 | 1.4 |
| USSR | 11.1 | 13.5 | 24.0 | *g* | *g* |
| Russia | *g* | *g* | *g* | 4.5 | 3.7 |
| South Korea | | | | 1.4 | 2.4 |

[a] Original source: www.cas.org [b] Modified from a table previously published in Chemistry: Decade by Decade, copyright 2007 by Arthur Greenberg, published by Facts On File, an imprint of Infobase Learning. Reprinted with permission of the publisher. [c] Peoples Republic of China (does not include Taiwan). [d] West Germany and East Germany percentages added together. [e] One total value for Germany. [f] British Commonwealth (including U.K., Canada, India, Australia, others) combined. [g] The USSR divided into individual countries after 1989. Only Russia data reported.

Recognition as an author or co-author of a scientific paper has always had its complexities and controversies. One aspect of this is the perception of "fairness" in listing authors (as well as the order of authors). Thus, Seeman and House (*2*) conducted a survey of academic chemists at Ph.D.-granting institutions. Of six hundred chemists who responded, some fifty percent claimed not to have received appropriate credit. Only a minority of these reported discussing the perceived offending behavior with the "offending" individual. The Seeman and House study focused on the asymmetry of relationships in the academic setting (student-faculty advisor and junior faculty-senior faculty among these). The reported small percentage of respondents who confronted the "offending" individual or sought the attention of the university was attributed to fear of retribution, being considered a trouble maker, or simply institutional indifference. The survey report recommended cultural changes in academic institutions, possibly including the introduction of departmental ombudsmen (*2*). However, these ambiguities have become considerably more complex in our current environment, due, in part, to the increasing role of highly interdisciplinary, multi-institutional science. It might be fair to say that half a century ago, in order to be an author or co-author on a paper, one would be expected to understand and be capable of answering relevant questions pertaining to most, if not all, of the paper. This would include full interpretation of NMR spectra, although not necessarily commanding the physics or being capable of tearing down and reassembling the instrument. Even then, an experimentalist employing quantum chemical calculations did not need to be a theoretician or be capable of writing computer code but most certainly needed to understand its applications and limitations and provide or cite some benchmarking (comparing agreement with experiment) at the level of theory employed.

In more recent decades, science studies have become far more interdisciplinary, as well as far larger. For example, the 2001 publication of the sequence of the human genome had over 200 co-authors from numerous institutions (*3*). Astrophysics and particle physics provide even more recent superb illustrations. The utterly spectacular observation on August 17, 2017, of the gravitational waves and electromagnetic observation of gamma-rays generated from a neutron star merger was surely one of the most important discoveries in the history of science. The report in *Physical Review Letters* listed a lead author and nearly 1200 co-authors from 162 listed institutions (*4*). However, the world's record holder appears to be the article in 2015 describing discovery of the long-sought Higgs boson (*5*). The paper in *Physical Review Letters* lists 5,154 authors and 344 research institutions. Although the present author should (perhaps) resist the temptation to exaggerate, in light of proliferating "bean counting" by hordes of administrators who have insisted on computerizing faculty annual reports and wish to turn faculty evaluations into simple numerical indices, the fraction 1/5,154 of a paper would not appear to improve one's case for a salary merit increment. Journal impact factor and especially news coverage might help a bit.

Returning to the issue of how much and how broad the knowledge of an individual author should be in a highly interdisciplinary multi-authored paper, I may cite an example from my own experience. During a significant period of my own career (*ca* 1980 through 1995), I was a participant in collaborative research in environmental chemistry. My research started as an exploration of carcinogenic polycyclic aromatic hydrocarbons (PAHs) on airborne particulate matter, their distribution (urban, industrial, suburban, rural sites) and seasonal variation, and quickly evolved into interdisciplinary projects involving sources, transport and reactivity in the troposphere, toxicity, short-term bioassays of toxicity, total human exposure, sampling strategies, and statistical analyses. The research included scientists at New Jersey Institute of Technology, New York University School of Medicine (Tuxedo, NY), Rutgers University, Robert Wood Johnson Medical School, New Jersey Department of Environmental Protection, Institute for Medical Research (Camden, NJ), and Columbia University School of Public Health. Thus, a paper or public presentation could include the work of organic chemists (my own work and that of my associates and students), toxicologists, bacteriologists, meteorologists, epidemiologists, and statisticians. Although I did learn aspects of Ames mutagenicity assays, Enzyme-linked immunosorbent assay (ELISA), air pollution wind roses, and some rudimentary statistical methods, it was impractical for me to answer a challenging question outside of details of, for example, benzo(a)pyrene reactivity or specifics of our high-performance liquid chromatography (HPLC) methodology. This was, most certainly a departure from the norms expected at the start of my scientific career.

Before moving on to questions of opportunities for authorship and fair assignment of authorship, it is appropriate to comment on some significant changes in the publications landscape. Some of these are very laudable, while others are downright sleazy. The increasing number of scientists and the continuous development of new fields such as nanotechnology and proteomics, coupled with increasing specialization within fields, has contributed to an amazing proliferation of journals. Some of the newer journals are excellent; some are inferior, as judged by low standards of review of submitted manuscripts; and some are simply money-making operations for unscrupulous publishers. Even publishing companies producing well-accepted journals continue to raise prices at rates far exceeding inflation. Some of the pressure on some larger institutional libraries has been mitigated through electronic subscriptions, which also relieves pressure on physical space in libraries. Some pushback by institutional subscribers and researchers has led to innovations such as the Public Library of Science (PLoS) which largely publishes in the life sciences and medicine. Initiated by a team of distinguished scientists in 2000 *PLoS One* and a series of other PLoS journals maintain high standards and public access. On the seamier side are a growing plethora of "predatory journals" and "Potemkin-village-style" professional meetings (*6*). Gina Kolata's *New York Times* article (*6*) focused in part on the World Academy of Science, Engineering and Technology (WASET). Having been an invitee to present at WASET conferences thanks to my "distinguished research etc, etc, etc" I visited WASET's website. There appear to be almost weekly meetings in world-class cities, but very modest local facilities. Deadlines for submitting manuscripts for review are but one month ahead of the meeting. While it is clear that colleges, universities and their scholars that understand the nature of research avoid these journals and meetings, pressures to "publish or perish" at less sophisticated institutions blessed with "bean-counting" administrators furnish the funding that keep such journals and meetings economically viable.

## Possible Solutions in Assigning Authorship in Scientific Publications

"Inappropriately assigning authorship credit" is most certainly professional misbehavior (*7*). This was one of some sixteen scientific misbehaviors identified by six focus groups involving 51 scientists from top-tier research universities. Ten misbehaviors were considered "most serious" with number one being "falsifying or 'cooking' research data". Four of six institutional research officers viewed any behaviors in this group of ten as "sanctionable". Authorship misbehavior was ranked number 12, just below the ten most serious. The authors of this study mailed 3,600 surveys to mid-career scientists (yielding 1,768 usable responses) and 4,160 surveys to early-career scientists (yielding 1,479 usable responses). It requested that the scientists self-report whether they had engaged in misbehaviors listed during the previous three years. On the authorship question 10.0% of the respondents reported their own (perceived) misbehavior. In the mid-career group (average age 44), 12.3% reported such behavior while the report in the early-career group (average age 35) was lower (7.4%). The study's authors suspect that the admissions of misconduct may be a bit on the low side due to reluctance to admit such behaviors. The early-career group is consistently lower in most, but not all, categories, possibly due to differences in opportunities to engage in such behaviors; risks associated with being caught at different professional stages; different education experiences and practices; or greater reluctance on the part of early-group researchers to report misbehaviors, as they may view themselves as particularly vulnerable to consequences. In any event, even 10% is a very significant percentage with reverberations well beyond the individual who commits "authorship misbehavior".

Among the clearest pathways to defining and understanding authorship is clarity in definitions and policy. A very cogent contribution is the report *Authorship in Scientific Publications. Analysis and Recommendations*, issued in 2013 by the Swiss Academies of Arts and Sciences (*8*). The report offers a general list of inclusive criteria for "determining entitlement to authorship":

- Making a substantial contribution to the planning, execution, evaluation and supervision of research;
- Involvement in writing the manuscript; and
- Approving the final version of the manuscript.

The Recommendations section of the report includes a brief and useful summary statement:

> "Anyone who, through his/her own scientific work, has made a substantial contribution to the planning, execution, evaluation or supervision of research, and to writing the manuscript, qualifies for authorship" (*8*).

It further indicates that occupying a managerial position does not in and of itself justify authorship, nor does being the manager who secured funding for the research. It noted further that listing authors in order of seniority is not appropriate.

The Swiss report provided a brief glossary related to publication including:

- Author with overall responsibility
- Corresponding author
- Ghostwriter
- Honorary authorship (gift authorship/guest authorship)
- Medical writer
- Scientific activities/scientific work
- Scientific seniority

The report is unambiguous in discouraging the co-listing of "honorary" ("gift" or "guest") authors and makes it clear that "ghost" writers (commissioned to write for another, usually for a fee) and "medical writers" (professional editors who "put scientific findings into a form suitable for publication") should not, if this is the extent of their activities, be listed as authors. To this, one should include the importance of a proper understanding of the criteria for authorship by technicians and graduate and undergraduate research students.

Significant attention is given in the Swiss report to the order of authors listed in the by-line of the paper. The most significant positions are first author and last author. Typically, but not always, one of these two authors is the "corresponding author". Tscharntke *et al* (*9*) published a paper that offered a taxonomy of schemes for determining author sequence, along with suggestions for preferred practices. The need for precise numbers or fractions is almost certainly driven by the "bean-counting" increasingly dominating higher education. Four approaches were identified:

1. The "sequence-determines-credit" approach (SDC). Order reflects declining (relative) contribution. In order to provide useful quantitative measures for evaluation (e.g. promotion and tenure, merit committees, etc.) these authors suggest whole impact factor for the first author, half impact for the second author, one-third impact for the third offer and continuation until the tenth order (one-tenth impact) and no further;
2. The "equal-contribution" norm (EC). Authors are listed in alphabetical order. This practice appears to be in decline (*10*). Tscharntke *et al* (*9*) suggest dividing the impact factor equally but no author receiving less than 5%.
3. The "first-last-author-emphasis" norm (FLAE). The suggestion is full impact to the first author, half impact to the last author, and the credit to the other authors is the impact divided by the total number of authors. (It is useful to note here that the relative importance of first versus last author does vary somewhat from discipline to discipline).
4. The "percent-contribution-indicated" approach (PCI). Tscharntke *et al* (*8*) note a growing trend for journals to detail each author's specific contribution, and they suggest this should further help to quantitate credit and aid any reader who wishes to pose questions to the relevant author.

Tscharntke *et al* (*9*) suggest that whatever approach is employed (*e.g.* 1, 2, 3, or 4 above), the authors of a paper should explicitly identify it in their paper. Beyond that as noted above, journals are increasingly suggesting or requiring that the roles of authors be identified. For example, in 2004, the editor of the *Proceedings of the National Academy of Science* suggested the following examples of attribution (*11*):

- Designed research
- Performed Research
- Contributed new reagents or analytic tools
- Analyzed data, or
- Wrote the paper

These are, of course, very broad and non-specific but do suggest an approach. An excellent editorial titled "Authorship matters" was published in *Nature Materials* in 2008, making the case for very careful consideration and standards in authorship (*12*). In 2017, there appeared an excellent set of definitions and guidelines: "*Nature* journals' authorship policy" (*13*). One particularly helpful set of guidelines pertains to explicit recognition of the responsibilities of the "corresponding author" pre- and post-publication.

## Fairness in Sequencing Authors

Issues in the sequence of authorships arise not only as matters of fairness in acknowledging credit but also because these are vital measures of professional standing and critical data in determinations of promotion, tenure, and merit in academic institutions. As Tscharntke *et al* (*9*) and countless other professionals acknowledge, the first and last places on multi-authored by-lines are the most prestigious. Although practices vary quite significantly by academic field, the first author is typically the person who has played the most major role (possibly performing the most experiments or contributing most to the study's conclusions). The last author may well be the senior author, who directs and supports the study. Here it must be noted that the professional expectation of this (last) author is significant contribution to the scientific content, not "honorary" authorship in accordance with merely providing funding or administrative leadership.

In recent years there has been particular and well-merited interest in the issue of gender in scholarly authorship (*14*). It is almost embarrassing to realize that the masterpiece *Conversations in Chemistry*, first published in London in 1806, was authored anonymously. This book passed through numerous editions and translations through 1853, and Edgar Fahs Smith estimated a total of 160,000 copies sold (*15*). Yet, the author, Jane (Haldimand) Marcet (1769-1858), only finally officially consented to be listed as author in the 1837 edition. Indeed, as a youth, Michael Faraday was first introduced to chemistry through that book, and he credited Mrs. Marcet as his first chemistry teacher (*15*). A reviewer of the present chapter pointed out that there are many uncertainties in counting all of an author's contributions. Authors who change their names due to marriage or divorce and authors with hyphenated names are particularly prone to uncertainty.

West et al (*14*) published the results of their survey of the enormous JSTOR digital library which includes the sciences and humanities from 1545 through the present. Launched in 1995, JSTOR is dedicated to making its digital library available easily and inexpensively. In their study, these researchers explored the trends in female authorships, including sole authorships, first and last authorships, and order of authorship in multi-authored papers. These were tabulated in different academic fields. The study demonstrates a dramatic rise in the percentage of PhDs in the physical sciences who are female. In the 1960s the percentage was 5% - 7% with a big jump in the 1980s and to 25% - 29% in the 2000s (*14*). This is reasonably consistent with a tally of 19.1% tenured and tenure-track female chemistry faculty in the 50 top research-funded American universities in the 2014-15 academic year (*16*). Some progress has been made in this area since 2001 where the percentage was only 10% (*16*). The leader in this department, Rutgers University, reports 33% women on its tenure/tenure-track faculty. But there remain other significant concerns (even conundrums) in the publishing arena. First, the point has been made that, while women spent considerably more "bench time" than their male colleagues in their first year of graduate school in experimental biology disciplines, the male students published more in their first year (*17*). That early publishing success was postulated to provide an early edge in competing for post-doctoral positions and assistant professorships (*17*). The origins of this disparity are not yet clear. Although many causes had been examined by others, Maggie Kuo suggests that laboratory dynamics (faculty/student) may be a significant if not the major issue (*17*). The article by West *et al* notes that in some fields (*e.g.* Cell & Molecular Biology), first authorships by women actually exceed their percentage of participation in authorship by-lines. However, they remain significantly under-represented in last authorships (*18*). This appears to be almost universally true in physics and mathematics (*18*).

Even as women are gaining ground in academia there remain glaring inconsistencies in recognition and rewards which may still lower their participation in high-flying careers in chemistry. In the lead editorial in an issue of *Chemical & Engineering News (C&EN)* focusing on women in chemistry ("Where are the Women?), a 2017 survey finds that 18.6% of positions on boards of directors in the chemical industry and 13.7% of executive officers of U.S. chemical companies are women (*19*). As low as these numbers are they do represent a roughly 50% increase over the previous decade. The same issue of *C&EN* notes that, of 175 chemistry Nobel winners, only four are women (Marie Curie, Irène Joliot-Curie, Dorothy Hodgkin, and Ada E. Yonath). Similarly, Huryn *et al* (*20*) noted that Gertrude B. Elion's Nobel Prize in Physiology and Medicine was only the fifth in this discipline to be awarded to a woman. That *C&EN* issue considers thirteen deserving women who did not receive a Nobel Prize in chemistry. What can be done to bring more women into tenure-track faculty positions in chemistry and other sciences and into other influential positions? Huryn *et al* (*20*) have noted the "leaky pipeline" that had women receiving more than half of the bachelors and doctoral degrees in science and engineering but comprising only 28% of employed scientists and engineers. These authors offer a series of recommendations to increase the representation of women in medicinal chemistry (*20*). This advice, summarized below, should be heeded in all fields of science and engineering.

- Be aware of the issue at your institution;
- Recognize differences in temperament, approach, and style that may be holding back advancements for women;
- Support the women who work in the scientific field. Provide invitations to give talks, attend get-togethers, and give credit where it is due;
- Elevate the visibility of excellent women scientists as role models;
- Support the creation of events, career advancement and affinity groups dedicated to mutual support;
- Identify senior leader sponsors to help support and promote excellent women;

- Have your institution take a strong stand and work toward gender parity at each level, including establishing a pro-active campaign to interview excellent female candidates for all positions (*20*).

As this chapter is being written, the United States and many other countries are confronting issues long festering that have been brought to light and publicized by the "Me Too" movement. In this regard, a timely article on sexual harassment in chemistry recently appeared in *C&EN* (*21*). The article raises the following alarming point concerning sexual harassment in academic chemistry departments: "It may be among the reasons women aren't reaching parity in chemistry Ph.D. programs and faculty positions." Clearly such violations of human decency must be confronted directly and treated fairly and decisively.

It is worthwhile to briefly reference the excellent commentary in *Nature* by Professor Carol V. Robinson, who achieved the honors of being the first female professor of chemistry at both the University of Cambridge and the University of Oxford (*22*). She too notes that the "greatest attrition occurs during the transition from PhD to research" (*22*). She cites a Royal Society of Chemistry survey that found the difficulties of combining family life and career as well as the dearth of role models to be major factors. Professor Robinson was inspired by Marie Curie. She and Nobel laureate Ada Yonath agree that academic jobs can also be family friendly. She cites Professor Yonath as saying that the demands of family and science can be balanced "as long as you have a passion for both" (*22*). In Professor Robinson's case, she actually took an eight-year career break to raise three children before returning to science. There, she took on a very unconventional project for a person studying structural biology. In spite of well-meaning advice from colleagues, she studied gas phase mass spectrometry of macromolecular complexes and has enjoyed a distinguished scientific career. At the risk of posing an impolite question at this juncture, one might ask: "How many Carol V. Robinsons and Ada Yonaths are there in the world?" A fair answer might be, "Many, many more than one might imagine!"

It would certainly be of interest to assess the authorship roles of under-represented minorities. This includes African Americans, Hispanics, Latinos/Latinas, Native Americans and multi-racial chemists and is surveyed by the Open Chemistry Collaborative in Diversity Equity (OXIDE) (*23*, *24*). However, the total percentage of chemistry faculty remains only slightly above 4% (1.6% African Americans; 2.8% Hispanic, Latinos, Latinas; <1% Native Americans and multi-racial) (*24*). In addition to the small numbers, simple scanning of author names will not necessarily provide these data. Indeed, some authors (and journals) appear to refer (or practice) using first (and middle) initials rather than first and middle names thus adding further complexity. Hence, it does not appear that there are yet truly reliable author surveys for underrepresented minorities in chemistry. Although colleges and universities are, for the most part, proactively working to increase racial and ethnic diversity, the pool of qualified applicants remains small. It is clear that more must be done to interest children at the elementary school and especially the middle school levels in science. Role models will play an increasingly important role.

I close with the thought that, up to and including the nineteenth century and a good part of the twentieth century, chemistry (and science in general) was explicitly recognized as a "man's game". At the National ACS Meeting in Washington, D.C., a symposium titled "Ladies in Waiting for Nobel Prizes. Overlooked Accomplishments of Women Chemists" addressed this issue (*25*). This was a follow-up to an earlier symposium on "Correcting the Errors and Oversights of the Nobel Prize Committee" (*26*). In retrospect, how unbelievably wasteful and reckless it was to exclude half of the talented and dedicated people on the planet. Years (hopefully not decades) from now, humankind will be amazed by the absence of participation relative to population by scientists we awkwardly group together as "underrepresented minorities." One can dramatically make this point by recognizing, during the 2017-2018 season by the Metropolitan Opera, the scintillating performance of Ms Pretty Yende as Adina in the staging of Donizetti's *L'Elisir d'Amore* and *Lucia di Lammermoor*. Ms Yende was born and came of age in post-Apartheid South Africa (*27*). Had she been born two decades earlier, it is not apparent that she would have ever had the opportunities that led her to the pinnacle of the opera world. Surely, the world would have been much poorer for it.

## References


1. National Academy of Science, National Academy of Engineering, and Institute of Medicine of the National Academies. *On Being a Scientist. A Guide to Responsible Conduct in Research*, 3rd ed.; The National Academy Press: Washington, DC, 2009, pp 35−38.
2. Seeman, J. I.; House, M. C. Authorship Issues and Conflict in the U.S. Academic Chemical Community. *Account. Res.* **2015**, *22*, 346–383.
3. Lander, E. S.; et al. International Human Genome Sequencing Consortium). Initial Sequence Analysis of the Human Genome. *Nature* **2001**, *409*, 860–921.
4. Abbot, B. P.; et al. (LIGO Scientific Collaboration, Virgo Collaboration). GW170817: Observation of Gravitational Waves from a Binary Neutron Star Inspiral. *Phys. Rev. Lett.* **2017**, *119*, 16101-1–161101-18.
5. Aad, G.; et al. (ATLAS Collaboration, CMS Collaboration). Combined Measurement of the Higgs Boson Mass in **pp** Collisions at $\sqrt{s}$ = 7 and 8 TeV with the ATLAS and CMS Experiments. *Phys. Rev. Lett.* **2015**, *114*, 191803-1–191803-33.
6. Kolata, G. In Academia, a Predatory Twist in Publishing. *N.Y. Times (N.Y. Ed.)* **2017**October31, D5.
7. Martinson, B. C.; Anderson, M. S.; de Vries, R. Scientists Behaving Badly. *Nature* **2005**, *435*, 737–738.
8. *English Translation Authorship in Scientific Publications. Analysis and Recommendations*; Brückner, C., Ed.; translation by Acheson, J. Scientific Integrity Committee of the Swiss Academies of Arts and Sciences: Bern, 2013. www.swiss-academies.ch/en/index/Publikationen/Richtlinien-Empfehlungen.html (accessed February 14, 2018).

9. Tscharntke, T.; Hochberg, M. E.; Rand, T. A.; Resh, V. H.; Krauss, J. Author Sequence and Credit for Contributions in Multiauthored Publication. *PLoS Biol.* **2007**, *5*, 13–14.
10. Waltman, L. An Empirical Analysis of the Use of Alphabetical Authorship Publishing. *J. Informetr.* **2012**, *6*, 700–711.
11. Cozzarelli, N. R. Responsible Authorship of Papers in PNAS. *Proc. Natl. Acad. Sci. U.S.A.* **2004**, *101*, 10495.
12. Editorial: Authorship Matters. *Nature Mater.* **2008**, *7*, 91.
13. Springer Nature. *Nature journals' Authorship Policy.* https://www.nature.com/authors/policies/authorship.html (accessed April 30, 2018).
14. West, J. D.; Jacquet, J.; King, M. M.; Correll, S. J.; Bergstrom, C. T. The Role of Gender in Scholarly Authorship. *PLoS One* **2013**, *8*, e66212.
15. Smith, E. F. *Old Chemistries*; McGraw-Hill: New York, 1927; pp 64−71.
16. Wang, L. Women Crack the Academic Glass Ceiling. *Chem. Eng. News.* **2016**, *94*, 18–21.
17. Kuo, M. *Women Miss Out on Authorship Opportunities Early On.* www.sciencemag.org/careers/2017/02/women-miss-out-authorship-opportunities-early (accessed April 30, 2018).
18. Pierson, E. In Science, It Matters That Women Come Last. *Five Thirty Eight.* August 5, 2014. https://fivethirtyeight.com/features/in-science-it-matters-that-women-come-in-last (accessed February 8, 2018.
19. Seijo, B. C. Where are the Women? *Chem. Eng. News.* **2017**, *95*, 2.
20. Huryn, D. M.; Bolognesi, M. L.; Young, W. B. Medicinal Chemistry: Where Are All The Women? *ACS Med. Chem. Lett.* **2017**, *8*, 900–902.
21. Wang, L.; Widener, A. Confronting Sexual Harassment in Chemistry. *Chem. Eng. News.* **2017**, *95*, 28–37.
22. Robinson, C. V. Women in Science. *Nature* **2011**, *476*, 273–275.
23. Wang, L.; Rovner, S. L. New Survey on Minority Chemistry Professors Released. *Chem. Eng. News.* **2015**, *93*, 37–39.
24. Widener, A. Few Gains for Minority Chemistry Professors. *Chem. Eng. News.* **2017**, *95*, 18–20.
25. Borman, S. These Female Scientists Should Have Won the Nobel. *Chem. Eng. News.* **2017**, *95*, 22–24.
26. *The Posthumous Nobel Prize in Chemistry. Volume 1. Correcting the Errors and Oversights of the Nobel Prize Committee*; Strom, E. T.; Mainz, V. V. Eds.; ACS Symposium Series 1262; American Chemical Society: Washington, DC, 2017.
27. Pretty Yende was born March 6, 1985 in Piet Retief, Mpumalanga, South Africa. She graduated from the South African College of Music and also from the *Accademia Teatro alla Scala* (Milan, Italy). She made her debut at the Metropolitan Opera in New York City on January 17, 2013. URL https://metopera/Discover/Artists/Sopranos/Pretty-Yende (accessed April 30, 2018).

## Chapter 16

# Case Study: Authorship in an Interdisciplinary World

**Jeffrey Kovac**

**Department of Chemistry, University of Tennessee, Knoxville, Tennessee 37996, United States**
***E-mail: jkovac@utk.edu.**

In an era of collaborative research when most published articles have multiple authors, assessing the relative contribution of individual contributors can be difficult but important. Scientists are judged by their peers in large part based on their intellectual contributions to research. In this case study, these issues come to a head in a difficult tenure and promotion case, where senior faculty disagree about the quality of the research of a junior colleague. The case raises several issues important in contemporary science including quantity vs. quality of research and the role of the journal impact factor in evaluating publications. This case will not be accessible to students who do not have experience with the evaluation of a research record.

## A Difficult Tenure and Promotion Case

John Campbell was being considered for promotion and tenure in the chemistry department at Midlevel State University. He thought he had a strong publication record, having coauthored twenty articles since his appointment. Almost all of his articles were collaborative. John was an infrared spectroscopist participating in several multi-university collaborations in which his spectroscopic capabilities were an important part of the materials chemistry research. John's name and his students' names appeared on all of the papers, but he was the lead or corresponding author on only two articles out of the twenty. Many of the articles had been published in journals with high Impact Factors, and several had been extensively cited. In the promotion and tenure dossier, John only identified his contribution in general terms, such as "project direction," so, it was difficult to determine how much of the intellectual contribution was his. John had good funding for his research, but almost all of his money came from being co-PI on grants that listed others as the lead PI, although he was the lead PI on one modest grant.

When the tenured faculty met to discuss John's promotion they were sharply divided. Those who supported his promotion pointed to the large number of publications and the fact that about half had appeared in high impact journals. They also noted that John had sufficient research funding to support his graduate students and postdocs and to maintain his program. Combined with satisfactory teaching and service, they felt that John's overall record met the standards of promotion.

About half the faculty disagreed. Their principal argument was that there was little evidence of intellectual leadership in the record. Although John and his research group had contributed significantly to a large number of papers, it seemed that the ideas for the projects had come from other collaborators. The fact that John was the lead PI on only one modest grant also suggested that, although he was a good collaborator, he did not have many original ideas, and it was good ideas that would sustain his career.

John's supporters countered by noting that collaborative, interdisciplinary research is increasingly important in science as the problems of interest become more and more complex. The days of articles with a single author or a faculty member plus a graduate student as authors are over. In the major journals, most articles have a half dozen or more authors. John had developed several excellent collaborations, for which his laboratory provided important data. In the contemporary research environment, John was a success.

After a long debate, a vote was taken by secret ballot and the motion to recommend John for tenure and promotion passed by a margin of ten to eight in favor of granting tenure. Because the vote was so close, it was decided that both a majority and minority report would be written and sent first to the department chair and then to the dean.

What do you think of John's record? Were the objections of the minority professionally justified? Does he deserve tenure and promotion?

Here are some questions to consider.

1. What criteria should be used in determining the contribution of individual researchers in a collaborative, interdisciplinary project? Should the area of research be considered when evaluating the relative contribution?
2. In assessing the record of an individual scientist, how important is a significant record of lead author (first author or last author) or corresponding author publications?
3. In assessing the research record of an individual scientist, how should quantity and quality of the publications be balanced?
4. What role, if any, should the journal Impact Factor have in assessing the quality of an individual publication?

## Chapter 17

# Case Study: Patterns in Authorship: Lessons in Diversity and Justice

**John D'Angelo***

**Division of Chemistry, Alfred University, Alfred, New York 14802, United States**
***E-mail: dangelo@alfred.edu.**

Authorship and how much a particular type (i.e patent, vs. books, vs. papers with collaborators, vs. papers without collaborators) of authorship is "worth" to a researcher's career is important to navigate. Authorship has well-established guidelines that are generally agreed upon. The standards for scholarship in tenure and promotion on the other hand vary from institution to institution and sometimes between departments within an institution. Herein, a hypothetical case that illustrates an issue uncovered by a recent study is presented. Students familiar with faculty rank and the tenure and promotion processes at US academic institutions are likely to benefit the most from this case study.

## Imagine This…

Organic chemists Francine Malot and Kevin Usa both start in their respective tenure-track positions in a Chemistry Division at a high-profile university at the same time. Thus, their tenure clocks are synchronized with each other (see the section titled "What Is Tenure and Why Is It Important?"). Tenure time comes, and they submit their portfolios. Both tenure candidates have been active in service serving on the same number of committees and both served as co-advisors for the graduate students' club. Both have been actively engaged in scholarship and have produced the same number of publications, though several of Francine's have more authors as her work is interdisciplinary and thereby involves more collaborators. When their tenure and promotion decisions are announced, Kevin is awarded tenure and promotion, while Francine is not recommended for tenure and promotion. The Committee cites their perception that Francine does not to have developed independence as a researcher. Francine files a grievance with the promotion and tenure committee and ultimately is granted an additional year to strengthen her tenure case. Upon being awarded tenure but not promotion the following year, she pursues a position at the University where her collaborator works and is immediately offered a position with both tenure and promotion to Associate Professor, an offer she accepts.

**Some Points To Ponder:**

- Did Francine behave unethically in leaving the University after being granted an extra year and ultimately tenure?
- Was the committee unethical in denying Francine tenure while granting tenure to Kevin, a candidate with virtually the same professional qualifications?

If you think that this sort of scenario is crazy, and that it is the 21st century and that this sort of thing doesn't happen anymore, you are sorely mistaken. Women, even today in 2018 still must fight for equal pay for equal work. Academia is not immune to this injustice. Faculty, especially in the sciences, are still predominantly male. And while in academia everyone likes to believe they are above discriminatory behavior, there is evidence that discrimination is still present. Such is the case in the next study, which can be used as an argument in favor of journals better articulating the relative contribution of authors. A so-called contributorship system may at a minimum make it harder for promotion and tenure committees to engage in such discriminatory evaluations of performance when considering raises and tenure or promotion decisions.

As part of her Ph.D. studies in economics at Harvard, Heather Sarsons (*1*) conducted a study on tenure decisions for economics professors. Her findings were disturbing, to put it mildly. Although it is not clear that the trends she found are applicable to other disciplines, that these trends **are** true in any academic discipline is...disappointing, at best. In her study, she evaluated the impact of co-authored manuscripts, as indicated in the academic economists' *curriculum vitae*, on their receipt of tenure. Her findings should concern everyone since she found that while sole-author papers counted the same in tenure decisions for men and women when the numbers of co-authored papers were compared (industry standard lists co-authors alphabetically), women were found to be disadvantaged and were less likely than men to receive tenure. The effect is particularly pronounced (against women) when the co-author is male. She also found in her study that in fields such as sociology where authors are listed in order of contribution and where author contribution is made clear, both men and women received equal credit for coauthored papers. That is, collaborative research does not negatively impact women. If this disparity is present in other fields (and we would be naïve to think that the sciences are immune), journals such as those published by the ACS would be just in clearly articulating the contributions of all authors for all the manuscripts they publish. Furthermore, that such a disparity is not as strong in fields that clearly articulate the contributions of authors should be a wake-up call to all fields that do not engage in this process. Even implying author contribution vis-à-vis relative author order may not suffice. If not for the social justice-inspired reasons, for the legal ones as it is discrimination to judge men and women unequally for the same work.

**Some Additional Points To Ponder:**

- Should it be responsibility of the tenure candidate to articulate to a committee the nature and extent of their contributions?
- Do the journals have an obligation to keep such a disparity in evaluation from happening, should they make it more clear because it best represents the contributions of multiple authors, or perhaps both or neither?
- Should collaborative work be worth less than non-collaborative work from a professional point of view?
- Is collaborative work necessarily an indication of lack of true independence?

## What Is Tenure and Why Is It Important?

Tenure, and whether it should exist, cease to exist, or be modified, is an interesting level of job security that exists almost exclusively in Academia. A discussion of how, if at all, tenure should exist is far beyond the present discussion, but the interested reader can find a wealth of information on the subject searching on the internet. Beware of what you read, however and be sure to read articles providing multiple perspectives, that of: faculty, administrators, and the public. To provide a better appreciation for the issues in this case, a discussion of tenure and its' benefits and impact on an individual's academic career may be helpful.

In higher education, although a tenured faculty member *could* be fired, the conferral of tenure on a faculty person amounts to job permanence, not dissimilar to that of a Supreme Court Justice. Once someone earns tenure (and it truly must be ***earned***), they can typically only lose their job because: their department or program is eliminated from the institution, and there is no other position they are willing to accept on campus; their institution has declared financial exigency (effectively, this is equivalent to a College or University declaring bankruptcy); or for just cause. What constitutes just cause may vary from institution to institution but generally this means committing a serious crime; gross, sustained incompetence; or extremely inappropriate behavior (e.g. coming to class inebriated, relationships with students).

In the ideal, tenure ensures academic and scholarly freedom. That is, it allows a faculty person to explore controversial pursuits without fear of losing their job. Although there are certainly some faculty, who abuse the system and either coast into the twilight of their careers or say inflammatory/incendiary things just because they feel like they can without penalty, tenure's continued existence is necessary to move scholarly work forward. Faculty persons must not work in fear of their work itself costing them their job. Tenure helps to prevent this. And this is not only true in the scholarly work venue. It also emboldens faculty members to experiment with their teaching methods more freely as they can avoid doing out of concern for low teaching evaluations, in fear of "this change in how I teach may get me fired if it doesn't work. Increasingly, the scholarship of teaching and learning is a recognized field of scholarly work. This increasing recognition of research into teaching methods is a trend that ought to continue as it will only improve teaching and thereby, education. However, teachers must feel free to take a chance on using new teaching methods. Such freedom may not be possible without tenure. Lastly, tenure also prevents age discrimination. an academic institution from saving money by purging all the experienced (and expensive) faculty persons and replacing them with younger, cheaper instructors.

## References

1. Sarsons, H. *Gender Differences in Recognition for Group Work.* https://scholar.harvard.edu/files/sarsons/files/full_v6.pdf (accessed May 11, 2018).

# Editors' Biographies

## Patricia Ann Mabrouk

Dr. Patricia Ann Mabrouk (PAM) is Professor of Chemistry & Chemical Biology at Northeastern University (NU). PAM received her A.B. in chemistry and mathematics from Wellesley College and her Ph.D. in inorganic chemistry at the Massachusetts Institute of Technology where she studied with Mark S. Wrighton. She received an NIH Postdoctoral Fellowship and studied with Edward Ira Solomon before joining the NU faculty in 1990. Over the years, PAM's research interests have included green polymer chemistry, biomolecular recognition in heme proteins, supercritical fluid electrochemistry, educational research on the efficacy of research ethics training, development of models for graduate student mentoring, problem-based learning, undergraduate research, research ethics, reflective journaling and graduate education.

PAM is an ACS Fellow, a CASE Massachusetts Professor of the Year, and an NSF CAREER awardee. Over the past twenty-seven years PAM has published over fifty peer-reviewed publications, obtained one U.S. patent, authored two books, served as Editor for an ACS Symposium Series volume on active learning, and organized or co-organized six symposia at ACS National Meetings since 2000. She is very active in ACS and CUR and has served in various capacities at the local and national levels. She is currently a member of the ACS Ethics Committee.

## Judith N. Currano

Judith N. Currano is the Head of the Chemistry Library at the University of Pennsylvania, where she teaches a graduate course in chemical information. She received her B.A. in chemistry and English from the University of Rochester, and she earned an M.S. in Library and Information Science at the University of Illinois at Urbana-Champaign. Her research interests lie in three areas of chemical information education: training chemists to navigate the literature, introducing basic chemical principles to librarians, and teaching substructure search techniques.

Judith received the Special Library Association's (SLA) 2016 Rose L. Vormelker Award in recognition of her commitment to chemical information literacy, a subject on which she regularly publishes and presents. She is active in SLA's Chemistry Division and offers short courses and Webinars that teach basic principles of chemistry and chemical information to librarians. Judith coedited "Chemical Information for Chemists: A Primer," published by the RSC in 2014, and coedited two ACS Symposium Series volumes the same year. She was elected to the Board of Trustees of the Cambridge Crystallographic Data Centre in 2014, and she has served on library advisory boards for many chemical information providers, including ACS Publications, CAS, *Science of Synthesis*, and Bio-Rad's KnowItAll. She is active in the ACS and serves in various capacities at national, division, and local levels. Judith currently chairs the ACS Committee on Ethics, and she has served as Chair of the ACS Division of Chemical Information (CINF) and has edited CINF's quarterly newsletter. In her spare time, Judith advises Penn's Women in Chemistry group, and she has been a Girl Scout troop leader since 1999.

# Indexes

# Author Index

# Subject Index

www.ingramcontent.com/pod-product-compliance
Ingram Content Group UK Ltd.
Pitfield, Milton Keynes, MK11 3LW, UK
UKHW020817290926
15470UKWH00003B/1222

9 780841 233393